Paleobiology of Giant Flightless Birds

Series Editor
André Mariotti

Paleobiology of Giant Flightless Birds

Delphine Angst
Eric Buffetaut

ISTE PRESS

ELSEVIER

First published 2017 in Great Britain and the United States by ISTE Press Ltd and Elsevier Ltd

ISTE Press Ltd
27-37 St George's Road
London SW19 4EU
UK

www.iste.co.uk

Elsevier Ltd
The Boulevard, Langford Lane
Kidlington, Oxford, OX5 1GB
UK

www.elsevier.com

Cover illustration by Agnès Angst. Reconstructions of the heads of (from left to right starting from the upper row): Phorusrhacidae, Aepyornithiformes, Gastornithidae, Dinornithiformes, Brontornithidae, Dromornithidae.

British Library Cataloguing-in-Publication Data
A CIP record for this book is available from the British Library
Library of Congress Cataloging in Publication Data
A catalog record for this book is available from the Library of Congress
ISBN 978-1-78548-136-9

Contents

Acknowledgments

First, we would like to thank all the scientific collaborators with whom we had the opportunity to work on the paleobiology and paleoecology of the large fossil terrestrial birds presented in this book, especially Romain Amiot, Christophe Lécuyer, François Fourel, Anick Abourachid, Anthony Herrel, Florent Goussard, Patricia Wills, Alexis Dollion, Anusuya Chinsamy-Turan and Aurore Canoville.

We also would like to thank the curators or associates of collections, Ronan Allain, Christine Argot, Christine Lefèvre, Sandra Chapman, Loïc Costeur, Yves Laurent, Emmanuel Robert, Didier Berthet, Daniel L. Brinkman, Marilyn Fox, Carl Mehling, Christopher Norris, Jacques Gauthier, Marcelo Reguero, Alejandro Kramarz, Federico Agnolin and Herculano Alvarenga for their help in accessing the specimens, which allowed us to complete part of the research presented in this book. And also thank Patrick and Annie Mechin for giving us access to part of the *Gargantuavis* specimens and François Escuillé for access to a metatarsus and a mandible of *Gastornis,* and the donation of the latter to the Natural History Museum of Toulouse. More generally, we thank the following institutions for giving us access to the specimens studied: Muséum National d'Histoire Naturelle (Paris), Université Claude Bernard Lyon 1 (Lyon), Musée des Confluences (Lyon), Natural History Museum (London), Naturhistorisches Museum (Basel), Muséum d'Histoire Naturelle de Toulouse (Toulouse), Museo de La Plata (La Plata), Museo Argentino de Ciencias Naturales Bernardino Rivadavia (Buenos Aires), American Museum of Natural History (New York) and Yale Peabody Museum (New Haven).

Also, a big thank you to the *Association Culturelle, Archéologique et Paléontologique de l'Ouest Biterrois* (Musée de Cruzy) and to its members for their help and support during the numerous paleontological excavation campaigns in Cruzy, which allowed us to excavate several *Gargantuavis* specimens, as well as to Stephen Giner, Myette Guiomar and the volunteer researchers who took part in the various exploration campaigns in the Var to look for *Gastornis* eggshells.

Finally, special thanks to our illustrator, Agnès Angst, who carried out all the bird reconstructions in this book, as well as the cover.

Preface

During bird evolution, some large-sized forms, unable to fly, have appeared several times from *Gargantuavis*, which lived during the Late Cretaceous, about 72 million years ago, to the Madagascan *Aepyornis* and the New Zealand moa, which only became extinct a few centuries ago. Such giant birds lived in several regions of the world, often but not only in an insular environment, and frequently played a significant role in the ecosystems they belonged to. According to a fairly widespread belief, some of them would have also competed with mammals for continental domination, at the beginning of the Cenozoic.

Far from forming a homogenous systematic whole, these birds belong to very different groups, several of which have no living representative. Even if in the past some authors tried to see these birds, which were a bit too quickly referred to as "running", as the result of a relatively simple evolutionary phenomenon linked to the loss of the ability to fly, further examination of their anatomical characteristics quickly reveals significant differences, reflecting adaptations to various ways of life. Beyond the fundamental physiological distinction between herbivores and carnivores (which in the case of forms with no living counterparts was not always easy to establish from fossils), we can reconstruct, in particular, different types of locomotion and diet, which can be associated with distinct ecologies, in different habitats. To reconstruct the paleobiology of these extinct organisms in all its various aspects, we must then use different approaches, some of which were only recently developed, from functional anatomy to isotopic geochemistry, finite element analysis and paleohistology. In this book, our goal was to collect the paleobiological information available on most of the groups of large flightless birds, known in the fossil record, in order to reconstruct as plausibly as possible their ways of life and their roles in ecosystems. We hope that the comparison of the results obtained can lead to a better understanding of the modalities of this recurrent evolutionary phenomenon, which led initially small flying beings to transform into purely terrestrial organisms reaching a considerable size and weight[1].

Delphine ANGST
Eric BUFFETAUT
August 2017

1 In this book, we took into account only very large extinct terrestrial birds, some of which with dimensions comparable to those of extant ratites (whose smallest is Darwin's rhea, *Pterocnemia pennata*, which can reach 1 m in height and weigh up to 28 kg, whereas the ostrich can exceed 2.50 m in height and 150 kg in weight), whereas others substantially exceeded them. We did not take into consideration other extinct flightless birds, such as the dodo (*Raphus cucullatus*) and the Rodrigues solitaire (*Pezophaps solitaria*), which can be considered as giants within their group (in this case, the Columbiformes), but had dimensions inferior to those of the very large birds studied in this book, and therefore raise different paleobiological issues. For the same reason, we did not discuss *Sylviornis*, an extinct large terrestrial bird from New Caledonia, belonging to the Galliformes. We also chose not to review some large-sized fossil forms that are not really fully known such as *Remiornis*, a possible ratite from the Upper Paleocene of Europe (Martin, 1992; Mayr, 2009), or the Eremopezidae from the Upper Eocene of Egypt, which were interpreted as very ancient representatives of the Aepyornithiformes (Lambrecht, 1933), but whose affinities remain unclear (Mayr, 2009), and on whose biology we have very little information.

1

General Introduction

1.1. Bird evolution

Our knowledge of bird evolution was incomplete for a long time, especially regarding the Mesozoic, because of a relative lack of fossils for significant periods of this group's history. Things have considerably improved over the last twenty years, in particular following the discovery of numerous Cretaceous birds, especially in China, but also in other regions of the world (South America, Europe, etc.). Contrary to what is sometimes claimed, bird fossils are not really rare and are even particularly abundant at localities with exceptional preservation (Lagerstätten), like, for example, the Lower Cretaceous lacustrine deposits in North-East China, or the Eocene Messel Pit, in Germany.

The fact that birds derive from small theropod dinosaurs, which already had feathers, is nowadays generally accepted. The stages going from gliding dinosaurs with two pairs of wings to birds capable of flapping flight, which is only performed by the upper limbs, are better known (Zheng *et al.*, 2013). It is now possible to trace back the evolutionary history of birds with some accuracy since the Late Jurassic and the famous *Archaeopteryx* (even though the exact phylogenetic position of the latter remains debated) (Mayr, 2016). A first adaptive radiation of birds, accompanied by a quick diversification, occurred as early as the Early Cretaceous. It consisted mainly of archaic forms, including in particular the Enantiornithes, which remained a very significant component of the avifaunas up to the end of the Cretaceous period. Modern types of birds (Neornithes) were present from the Late Cretaceous onward, but most of their adaptive radiation took place during the Cenozoic, after the mass extinction at the Cretaceous–Paleogene boundary, which seems to have caused the disappearance of the archaic groups (Feduccia, 2014). The current great diversity of birds (approximately 10000 species) is the result of this radiation, which happened at the same time as that of the mammals.

It is in this context that the appearance, on several occasions, of very large birds unable to fly took place, the oldest case known being *Gargantuavis* during the Late Cretaceous. The paleobiology of these extinct giant flightless birds, ranging in time from the Cretaceous to the Holocene, is the subject-matter of this book.

1.2. Extant large terrestrial birds

There is currently a certain number of large birds unable to fly, belonging to the group known as Palaeognathae (Krumbiegel, 1966; Folch, 1992), which is defined on the basis of the structure of the palate, as opposed to the Neognathae (the group including most of the extant birds). These flightless paleognaths form the ratite group, which also includes fossil forms such as the New Zealand Dinornithiformes and the Madagascan Aepyornithiformes. The other extant paleognaths are the tinamous from South America, which are able to fly. However, recent molecular and genomic studies suggest that tinamous are the sister-group of the Dinornithiformes (Baker *et al.*, 2014), which implies that ratites are a paraphyletic group and that the loss of the ability to fly occurred on several occasions among paleognaths, followed by convergent evolution phenomena.

Extant large running birds can be divided into four families:

– Struthionidae, with the genus *Struthio* (ostriches), exclusively African today, but which also lived in Asia until recent times, and earlier in Europe.

– Rheidae, with the genera *Rhea* and *Pterocnemia* (the latter is not always differentiated from *Rhea*). They are the South American rheas.

– Casuariidae, with the genus *Casuarius*, which includes the cassowaries of New Guinea and Australia.

– Dromaiidae, with the genus *Dromaius*, the Australian emu.

All these birds are large, the largest being the ostrich (*Struthio camelus*), which can reach a height of 2.75 m and a weight of 150 kg, and the smallest being Darwin's rhea (*Pterocnemia pennata*) with a height of 1 m and a weight of 28 kg. Extant ratites have a cursorial mode of locomotion, which is reflected by the proportions of their long and narrow metatarsus (Angst *et al.*, 2016). Most of them live in open habitats, which are favorable to fast running, except for cassowaries, which live in a forest habitat. The biology of extant ratites is rather well known, especially since some species (in particular the ostrich and the emu) are bred in captivity. We thus have numerous data about their physiology, diet, reproduction, locomotion, behavior and habitat. This information can be useful for the paleobiological reconstructions of extinct giant birds. However, we must exercise caution in this field, because while some of these extinct large birds (Aepyornithiformes, Dinornithiformes) were paleognaths more or less related to modern ratites, others (Gastornithidae, Dromornithidae, Phorusrhacidae) were neognaths belonging to various groups very different from the ratites. In addition, the examination of the anatomy of some giant fossil birds reveals a graviportal mode of locomotion different from the cursorial one of the extant ratites. Therefore, it is erroneous to systematically refer to these extinct forms as "running birds" to indicate that they were unable to fly, as is sometimes done. Finally, while extant ratites are all herbivores, some extinct giant birds (Phorusrhacidae) were clearly carnivores. A detailed paleobiological study, based on the paleontological data available, is then required for each group, if we want to try to convincingly reconstruct the way of life of these birds.

1.3. General stratigraphic distribution

Fossil and subfossil large terrestrial birds are known from the Late Cretaceous to the Holocene (Figure 1.1). The oldest correspond to *Gargantuavis*, from the Late Cretaceous of southern France and northern Spain (Buffetaut *et al.*, 1995; Buffetaut & Le Lœuff, 1998; Buffetaut *et al.*, 2015; Buffetaut & Angst, 2016a-b). We then have the Gastornithidae, which are the first large terrestrial birds known in the Tertiary. These birds are present from the mid-Paleocene to the mid-Eocene in Europe, and from the Early Eocene in North America and in Asia (Andors, 1995; Buffetaut & Angst, 2014a-b). Similarly, the family Phorusrhacidae, also called "Terror birds", was predominantly present in South America from possibly the mid-Paleocene (their presence in the Paleocene of Brazil is disputed) to the end of the Pliocene or possibly the Pleistocene (Alvarenga & Höfling, 2003; Alvarenga *et al.*, 2011). They also occurred in North America during the Pliocene (Alvarenga & Höfling, 2003; Alvarenga *et al.*, 2011) in North Africa during the Early Eocene (Mourer-Chauviré *et al.*, 2011) and in Europe during the mid-Eocene (Lutetian) (Angst *et al.*, 2013). At the same time, the family Brontornithidae, initially included within the Phorusrhacidae and now considered as a separate group on the basis of different anatomical studies (Agnolin, 2007, Mayr, 2009; Agnolin, 2013; Buffetaut, 2014), is known from the Miocene and Oligocene of South America (Agnolin, 2007; Buffetaut, 2016). The Dromornithidae are known in Australia from the end of the Oligocene to the arrival of the first *Homo sapiens* during the Pleistocene. Finally, two other subfossil groups coexisted with humans during at least part of their history, and became extinct because of human intervention. The Dinornithiformes, more commonly called moa and endemic to New Zealand, are known as early as the Miocene and disappeared shortly after the arrival of the first humans in the 13th Century A.D. (Allentoft *et al.*, 2014). The order Aepyornithiformes, endemic to Madagascar, is known from the Pleistocene to the recent Holocene (Clarke *et al.*, 2006).

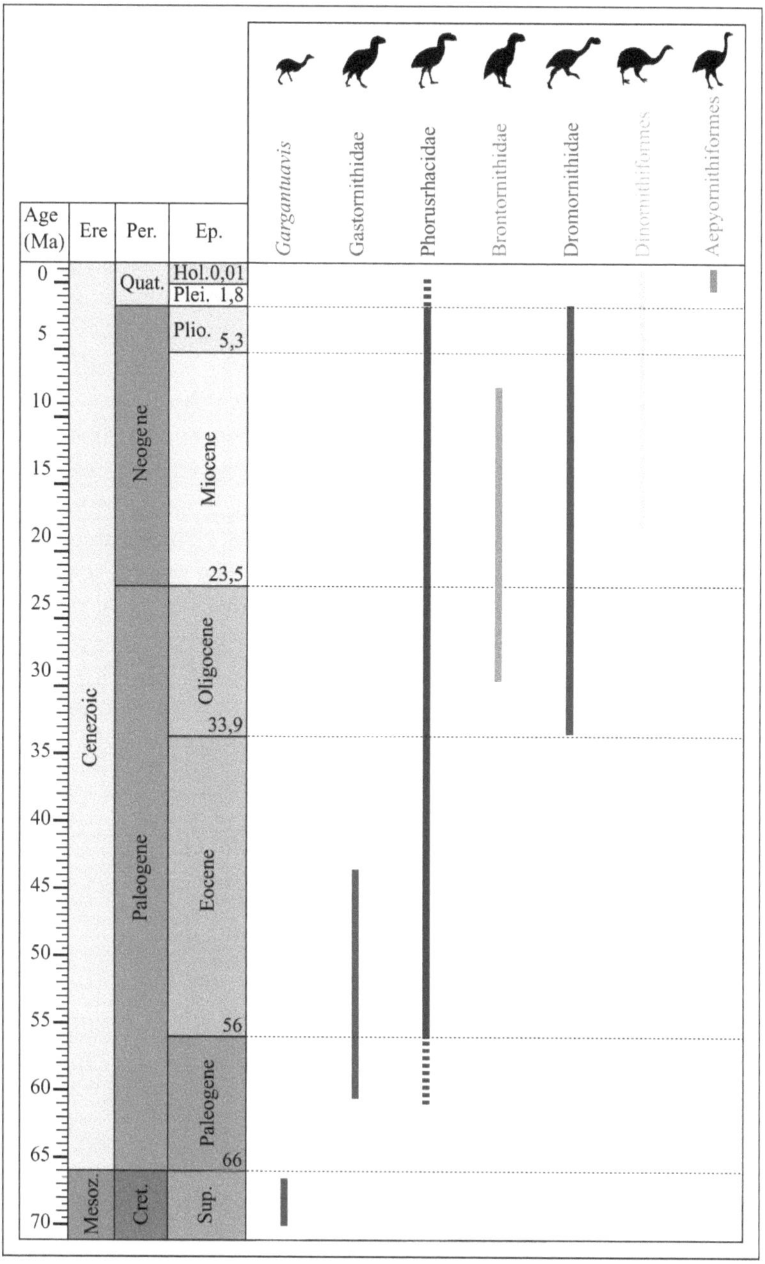

Figure 1.1. *Stratigraphic distribution of the large fossil and subfossil terrestrial birds. For a color version of the figure, see www.iste.co.uk/angst/birds.zip*

1.4. General geographical distribution

Some of these fossil and subfossil large terrestrial birds had a very wide geographical distribution spreading over several continents (Figure 1.2). It is especially the case of the Gastornithidae, which are known not only from Europe (France, Germany, Belgium, England) (Martin, 1992; Buffetaut & Angst, 2014a) but also from North America (USA, Canada) (Andors, 1988, 1992) and from Asia (China) (Hou, 1980). A second group with a very wide geographical distribution corresponds to the Phorusrhacidae, which are known not only from South America (Argentina, Uruguay, Brazil) (Alvarenga & Höfling, 2003; Alvarenga *et al.*, 2011), and North America (USA) (Alvarenga & Höfling, 2003; Alvarenga *et al.*, 2011), but also from North Africa (Algeria) (Mourer-Chauviré *et al.*, 2011) and Europe (France, Switzerland) (Angst *et al.*, 2013). However, the distribution of other fossil groups is limited to a continent, a country or a region. Thus, *Gargantuavis* is known from southern France and northern Spain (Buffetaut & Angst, 2016a), the Brontornithidae are known from South America (Argentina, Bolivia) (Alvarenga & Höfling, 2003; Alvarenga *et al.*, 2011; Buffetaut, 2016), the Dinornithiformes from New Zealand (Worthy & Holdaway, 2002) and the Dromornithidae from Australia (Murray & Vickers-Rich, 2004). Finally, the Aepyornithiformes are only known from the island of Madagascar (Hume & Walters, 2012).

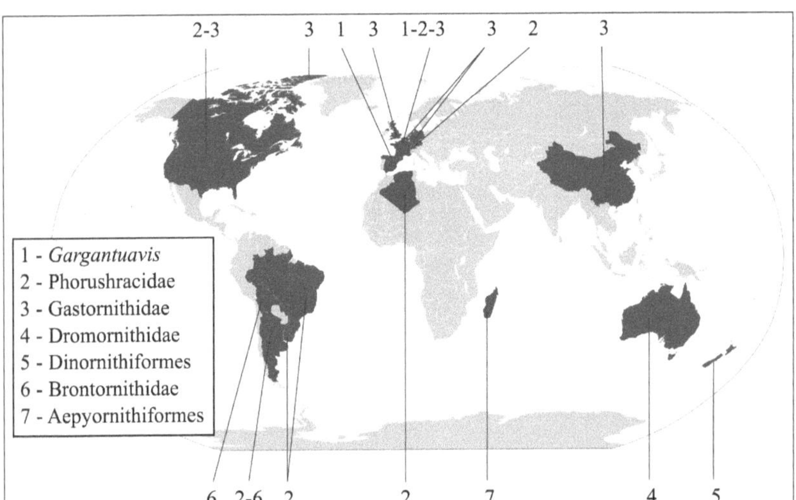

Figure 1.2. *Geographical distribution of large fossil and subfossil birds mentioned in text. For a color version of the figure, see www.iste.co.uk/angst/birds.zip*

1.5. Phylogenetic position of large terrestrial fossil birds

The giant birds that are the subject of this book belong to groups with very different positions in bird phylogeny (Figure 1.3). Their shared characters are a large size and the loss of the ability to fly (indicated by wing reduction), which are the result of convergent evolutions.

We can sum up the phylogenetic positions of the groups taken into consideration as follows:

– *Gargantuavis*: the systematic position of this giant bird from the Late Cretaceous of Europe is uncertain, because the available osteological material remains limited. It seems that it can be placed among the basal Ornithurae (Buffetaut & Angst 2016a).

– Dinornithiformes and Aepyornithiformes: these two groups belong to the Palaeognathae and are traditionally placed among ratites (flightless paleognaths). According to recent molecular and genomic data, the Dinornithiformes are the sister-group of tinamous (flying paleognaths), which implies that ratites are paraphyletic and that the loss of flight intervened several times. The Madagascan Aepyornithiformes may have New Zealand kiwis as sister-group. The phylogenetic history of ratites seems complex, and numerous aspects remain unclear (especially regarding their paleobiogeography).

– Gastornithidae: Andors' opinion (1992), who considers the Gastornithidae as rather basal Anseriformes, seems widely accepted.

– Brontornithidae: these South American giant birds have long been considered as very closely related to the Phorusrhacidae, or simply as robust large-sized phorusrhacids. Agnolin's opinion (2007), who considers them as basal Anseriformes, is not unanimously accepted, as some authors still favor close relationships with the Phorusrhacidae (Alvarenga *et al.*, 2011). Agnolin's classification, based on strong evidence, is nevertheless widely accepted and followed in this book.

– Dromornithidae: after having been considered for a long time as ratites, the Dromornithidae were considered as Anseriformes (Murray & Vickers-Rich, 2004), and this opinion is generally accepted nowadays.

– Phorusrhacidae: various opinions were expressed regarding the phylogenetic affinities of these large carnivorous birds. A rather close kinship with extant seriemas (Cariamidae) is generally accepted today, and the Phorusrhacidae can be placed with the Cariamidae in the order Cariamiformes.

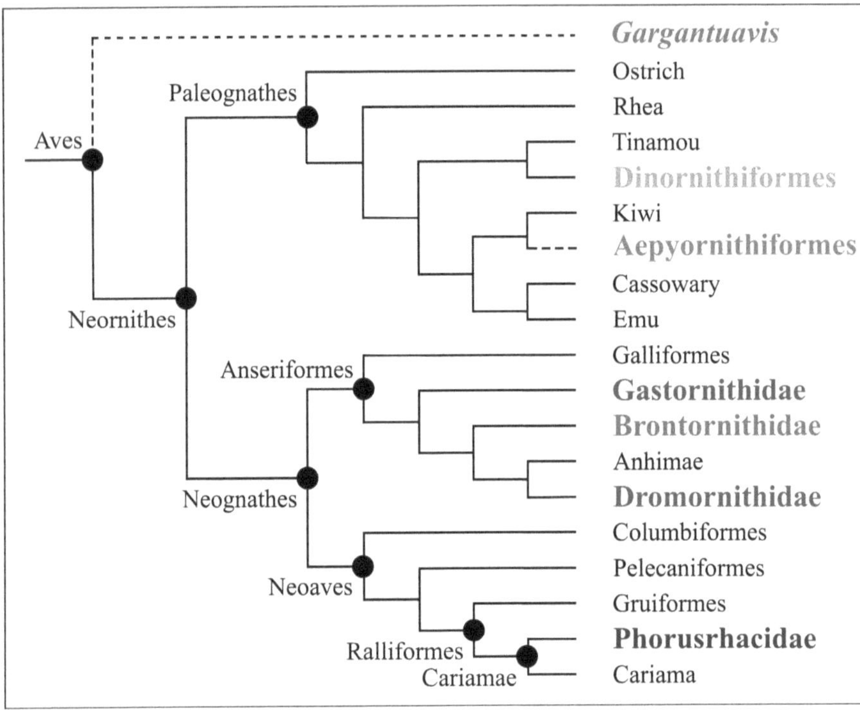

Figure 1.3. *Simplified phylogeny of the large fossil and subfossil terrestrial birds. For a color version of the figure, see www.iste.co.uk/angst/birds.zip*

1.6. Methodological approaches used

This part is not designed to draw an exhaustive list of the methods used in paleontology, but to explain the various approaches that will be later mentioned in this book to study the paleobiology and paleoecology of fossil terrestrial birds. These methods include the use of equations based on living organisms transposed into the fossil record, isotope geochemistry, functional morphology, finite element analysis (FEA), bone histology and finally bird anatomy in general.

1.6.1. *Equations derived from living organisms*

The paleobiological study of fossil vertebrates, and especially birds, frequently uses quantitative models based on simple equations. These equations are first established for living organisms whose biology is well known, and then, these models are transposed into the fossil record in order to determine different factors. Within the framework of this book, different parameters were studied using this approach. The use of equations allows us to estimate *inter alia* the body mass of birds, their type of locomotion, the size of eggs, the mass of these eggs and even the body mass of the female which laid these eggs, or the incubation duration.

1.6.1.1. *Body mass*

Several quantitative methods to estimate the body mass of birds were proposed during the 1940s. One of the first attempts at body mass estimation was proposed by Amadon (1947). He offered a correlation between the mass of large birds and body length (measured between the anterior edge of the thorax and the center of the acetabulum) and the surface area of a cross-section of the femur at the thinnest level of the diaphysis. However, in both cases, these measurements are difficult to get from fossils. Indeed, the measurement of the body length requires a complete articulated specimen in an appropriate position for measurement. Similarly, to measure the area of the femur section, we need either a natural break at the right place, or to section the specimen, or to estimate this area by considering the section as circular. However, such options are not always available, or they inevitably cause significant measurement errors that will have an impact on the bird mass estimation.

Later, Alexander (1983) proposed estimating mass from body volume estimations. To this end, a life-size model or a scale model of the organism studied must be built and immersed in water in order to estimate its volume from the volume of water displaced. This volume is then converted into mass by multiplying it by the density of a goose (measured by means of a goose carcass and its volume in water, while considering water density as 1). However, this method poses several problems. First, it requires the creation of a model that must be as accurate as possible. However, this stage is a potential source of numerous errors, and might include in the model a very significant number of uncertainties. Indeed, in order to create this model, particularly well-preserved fossil remains of the organism are needed, such as a complete skeleton, which is not available for many fossil groups. In addition, even in cases where the fossil record is exceptional, the creation of the model requires the reconstruction of the animal's soft tissues, including in particular muscles, fat and skin. Even though in some cases, muscular insertion areas are visible (which is not always the case), they make it possible in the best-case scenario to quantitatively know whether the muscle is large, but no method currently allows us to know exactly what the volume of these muscles was. This method then depends exclusively on the reconstruction of the organism, which is technically and scientifically difficult to achieve in a sufficiently accurate manner.

A third estimation method of the mass of an organism was proposed by Hurlburt (1999) under the name of GDI (graphic double integration). This method is based on the cutting of the animal's body into a set of elliptic cylinders, for which both perpendicular diameters were measured. The volume of these cylinders is then computable, and by considering water density as 1, the volume (in ml) is then equivalent to the mass (in kg). The measurement of both cylinder diameters can be performed either directly on the specimen, or via photos or illustrations, which makes measurements far simpler to take. However again, this method requires a complete fossil organism with either preservation or a reconstruction of soft tissue, which is very rare in the first case, and might add a great number of uncertainties in the second case.

Other methods offer to use the measurements of the hindlimb bones to estimate body mass. After Amadon (1947), the first to have used this approach were Campbell and Tonni (1983). They proposed an equation linking body mass to the minimum circumference of the tibiotarsal diaphysis. However, the model they offer is established from the study of extant flying birds with a mass inferior to 80 kg, which is not the case for the vast majority of birds studied in this book, making the application of this method difficult (as the birds studied in this book fall outside the framework of the model application). Later, Anderson *et al.* (1985) proposed a similar model estimating body mass from the femur circumference, measured at the level of half the maximum length of the diaphysis. However, here again, they built their model with an ostrich weighing 80.9 kg as the heaviest bird, which also makes this model inapplicable to some of the larger birds.

Finally, the last method to estimate the body mass of fossil birds is the one proposed by Campbell and Marcus (1992), by means of the minimum circumference of the femur and tibiotarsus. This method has several advantages. First, it is established over a very wide range of extant birds, including ostriches weighing 138 and 161 kg, which correspond to the highest possible weights for extant terrestrial birds. In addition, it uses the minimum circumference, which is an easy measurement to take, even if the bone is not fully complete, which is impossible when the circumference at half the diaphysis length is used. This then increases the number of possible measurements on fossil material, because bone ends do not need to be complete. This method is therefore the most suitable to estimate the body mass of fossil large terrestrial birds. Campbell and Marcus (1992) offer two equations linking the body mass (M) in kilograms to the minimum circumference of the femoral (LC_F) [1.1] or the tibiotarsal (LC_T) diaphysis [1.2] in millimeters. It should however be noted that equation [1.1] often provides body mass estimations higher than equation [1.2].

$$\mathrm{M} = (10^{2.411\times\log_{10}(\mathrm{LC_F})-0.065})/1000 \quad [1.1]$$

$$\mathrm{M} = (10^{2.424\times\log_{10}(\mathrm{LC_T})+0.076})/1000 \quad [1.2]$$

Body mass can also be estimated through another approach using eggs. There are different equations proposed by Dyke and Kaiser (2010), making it possible to link the mass of an egg (M_{egg}) to the body mass of the female (M_{body}) that laid it. There are two different equations according to the juveniles' mode of development. For the nidicolous (altricial) bird type, that is, when parents take care of the young after birth, equation [1.3] must be used. However, for the nidifugous (precocial) bird type, that is, when the young are independent soon after birth, equation [1.4] must be used.

$$M_{egg} = -0.659394 + 0.7889097 \times M_{body} \quad [1.3]$$

$$M_{egg} = -0.164615 + 0.6451872 \times M_{body} \quad [1.4]$$

1.6.1.2. *Locomotion type*

Most often, in the fossil record, the locomotion category (terrestrial, aquatic, semi-aquatic or aerial) is generally relatively easy to determine. However, getting more detailed information regarding the overall walking speed or the endurance capacity often remains difficult to determine. In order to better frame these concepts of locomotion in a terrestrial context, two main types of locomotion on land were defined, first for mammals, then extrapolated to terrestrial birds (Gregory, 1912; Osborn, 1934; Smith & Savage, 1956; Coombs Jr, 1978; Stein & Casinos, 1997; Carrano, 1999; Murray & Vickers-Rich, 2004), between which all intermediate cases are possible. The cursorial locomotion type corresponds to a rapid movement that can last for a long time, as is the case, for example, of extant ostriches. In contrast, the graviportal type of locomotion was defined for terrestrial animals with column-like limbs and with a relatively slow locomotion, with an ability to occasionally run. Even though this leg bone organization is not found among birds with a slow locomotion, the term graviportal is also commonly used for terrestrial birds, which is, for example, the case of the Aepyornithiformes (as no extant terrestrial bird uses this locomotion type anymore) (Amadon, 1947; Rasmussen *et al.*, 2001; Murray & Vickers-Rich, 2004). Determining the type of locomotion of terrestrial birds and, in particular, fossil terrestrial birds, has been done since the 1960s by means of the method proposed by Storer (1960). This method uses the proportions of the maximum length of the three hindlimb bones of terrestrial birds (femur, tibiotarsus and tarsometatarsus, Figure 1.22). The relative proportion of the tarsometatarsus compared with the two other bones makes it possible to estimate the locomotion type. Indeed, a bird with a cursorial type of locomotion will have a long tarsometatarsus, corresponding to a length globally superior to 30% of the total length of the hindlimb. Conversely, graviportal birds will have a shorter tarsometatarsus, the length of which will be inferior to 30% of the total length of the hindlimb (Figure 1.4).

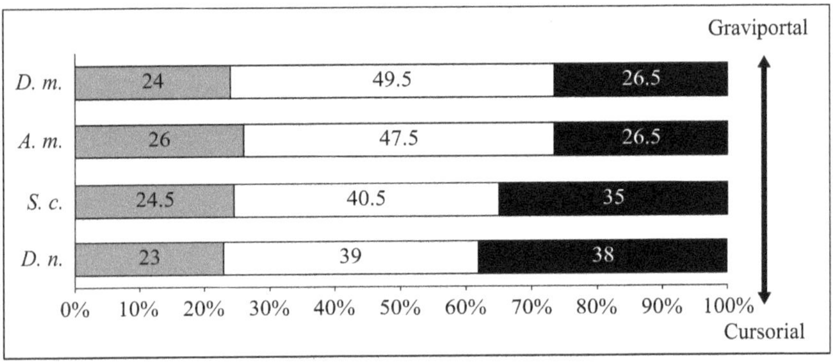

Figure 1.4. *Estimation of the locomotion type through Storer's method (1960).* D. m.: Dinornis maximus, A. m.: Aepyornis medius, S. c.: Struthio camelus, D. n.: Dromaius novaehollandiae

However, this model poses several problems. First, it requires complete bones in order to be able to measure the maximum length, which is not always possible in the fossil record. In addition, this model requires three bones of the same leg from the same individual. However, in the case of fossil birds, it is extremely rare to have enough complete articulated specimens to be able to use this method. It is then often necessary to take length measurements from several different individuals and to calculate averages in order to be able to use Storer's model. However, calculating the average length for each bone smoothes values, and so makes this method less reliable. Finally, for some of the terrestrial birds with an intermediate type of locomotion (e.g. cassowaries), this model does not have enough resolution to identify the locomotion type.

In order to solve these problems, a new locomotion estimation model for terrestrial birds has recently been proposed (Angst *et al.*, 2016). This new method is based on the maximum length-to-minimum width ratio of the tarsometatarsus, which is the bone the most sensitive to the mechanical constraints linked to locomotion, and which is already used in Storer's method (1960). The length is measured from the point of the most protruding process of the proximal end, up to the point of the most protruding trochlea of the distal portion (Figure 1.5). The minimum width is measured in dorsal or ventral view, at the narrowest level of the diaphysis.

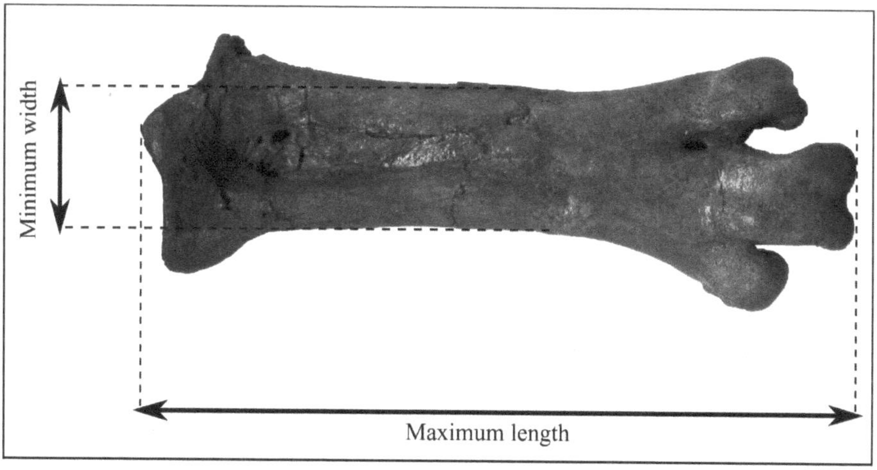

Figure 1.5. *Location of the measurements on the tarsometatarsus according to Angst* et al. *(2016). For a color version of the figure, see www.iste.co.uk/angst/birds.zip*

Thanks to a model established from extant or subfossil birds whose locomotion type is known, it could be shown that the length-to-width ratio of the tarsometatarsus of cursorial birds is superior to 12, whereas the ratio for graviportal birds is inferior to 12 (Figure 1.6). The former have a very narrow and slender tarsometatarsus, and the latter a very wide and massive tarsometatarsus. Therefore, this method not only allows us to differentiate both extreme cases, but its really good resolution also makes it possible to identify the locomotion type of all the intermediate cases not easily interpreted with Storer's model. This method thus has a high resolution, but can also very easily be applied to the fossil record, because it only requires two simple measurements (maximum length, minimum width) from a single bone (tarsometatarsus).

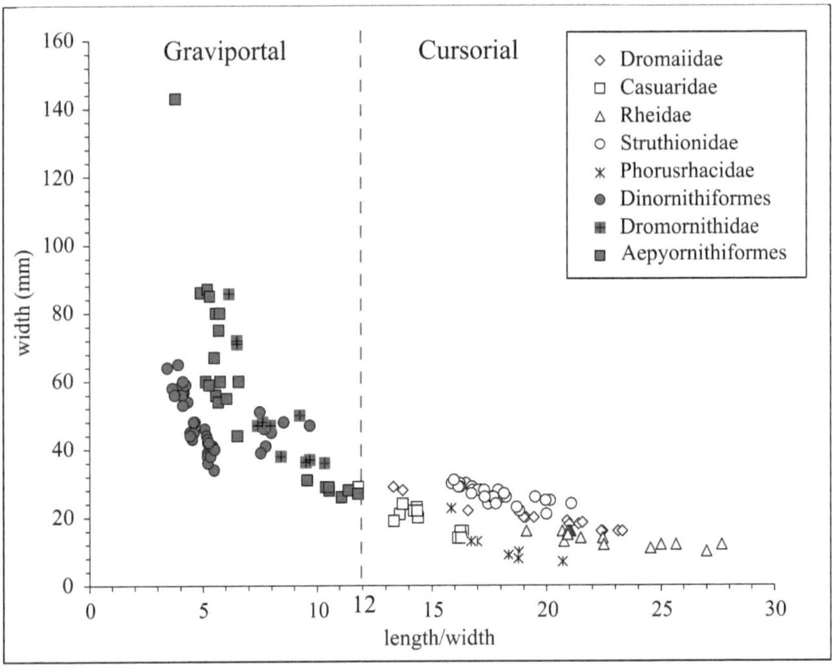

Figure 1.6. *Estimation of the locomotion type through the method of Angst* et al. *(2016). For a color version of the figure, see www.iste.co.uk/angst/birds.zip*

1.6.1.3. *Estimation of walking speed by means of footprints*

It is possible to estimate the walking speed from trackways using the model proposed by Alexander (1989). After studying a number of tracks from different terrestrial mammals, this author proposed an equation making it possible initially to estimate the walking speed of dinosaurs, and also applicable to large terrestrial birds. The idea is that the faster an animal walks, and the larger an animal is, the longer its stride is (with a stride corresponding to the distance between a footprint and the following footprint of the same leg). This can then be summed up by equation [1.5], which allows us to link the relative length of the stride (L. relative stride) to the length of the hindlimb (L. h.lim., corresponding to the height at the hip when the animal is standing idle), and which starts from the premise that whatever the animal, at equal speed and for legs with equivalent length, the stride will be the same. Speed itself is also estimated by means of equation [1.6] (with the gravitational acceleration (g.a.) = 10 m.s^2), and represented by the dimensionless speed, which makes it possible to eliminate the need for the size of the animal

$$L.relative\ stride = \frac{L.\ stride}{L.\ h.\ lim.} \qquad [1.5]$$

$$dimensionless\ speed = \frac{speed}{\sqrt{(L.\ h.\ lim. \times g.a.)}} \qquad [1.6]$$

From these two equations, Alexander succeeded in building a graph (Figure 1.7) which makes it possible to estimate dimensionless speed from the relative length of the stride, which is then valid for all animals whatever their size, and whatever the hardness of the substrate on which the footprints were left.

Thus, from the stride length and the hindlimb length of the animal that left its footprints, it is then possible to estimate its dimensionless speed by means of graph (1.9), and then the walking speed itself by means of equation [1.6].

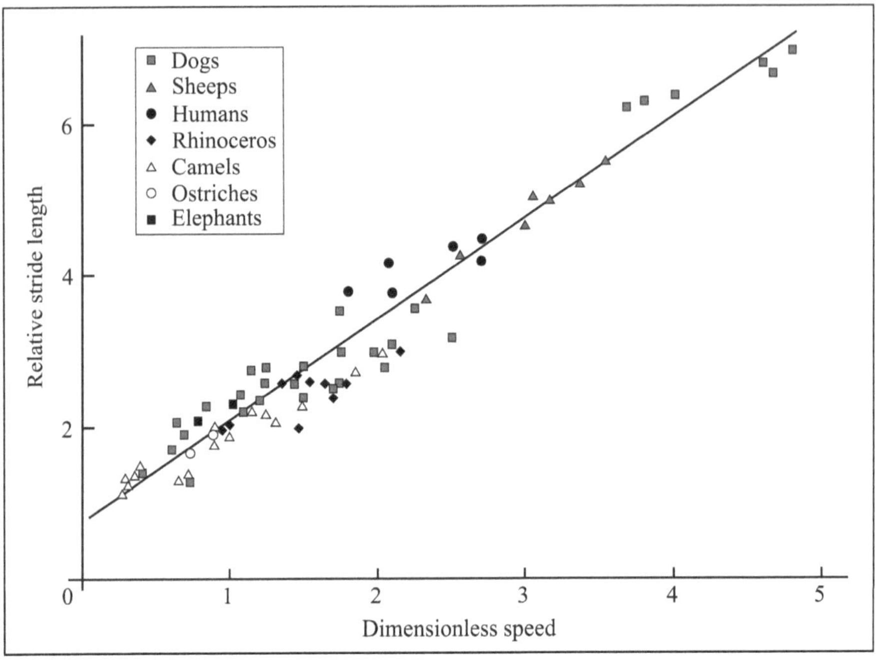

Figure 1.7. *Graph of dimensionless speed according to the relative stride length for various terrestrial vertebrates. From Alexander (1989). For a color version of the figure, see www.iste.co.uk/angst/birds.zip*

1.6.1.4. *Size, volume, mass and incubation duration of fossil eggs*

1.6.1.4.1. Size of fossil eggs

The estimation of the size of fossil birds' eggs is possible even from egg fragments, thanks to the use of equations established from the study of extant birds' eggs. Indeed, Williams (1981) proposed measuring the curvature of eggshell fragments in all the spatial directions using a spherometer (Figure 1.8(A)). This then allows us to estimate the value of the maximum curvature (C_W) and minimum curvature (C_L), which respectively correspond to the transversal curvature and the longitudinal curvature of the egg (Figure 1.8(B)). As a spherometer is an optical measurement instrument, values are therefore obtained in diopters and should be converted into millimeters following equation [1.7]:

$$D = \left(\frac{1,523 - 1}{curvature} \times 1000 \right) \times 2 \qquad [1.7]$$

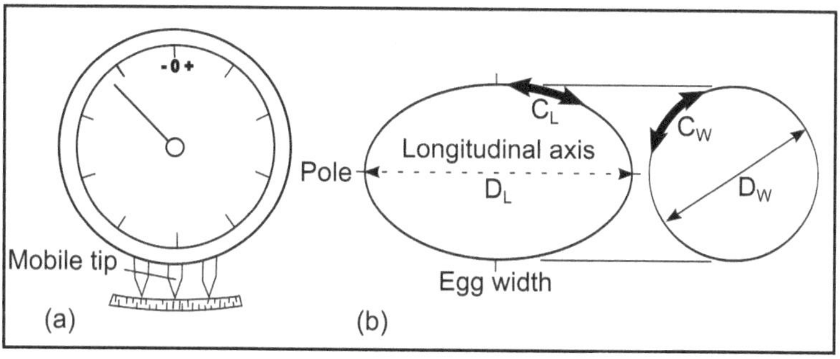

Figure 1.8. *Explanation of egg measurements according to Williams (1981). (A) Measurement of a shell curvature with a spherometer. (B) Egg geometry: C_W is the transversal curvature and C_L is the longitudinal curvature; D_W is the transversal diameter and D_L the longitudinal diameter*

The maximum curvature (C_W) thus provides the minimum diameter of the egg (D_W), which corresponds to the transversal diameter and the width of the egg, if we consider that the transversal section of the egg is circular. The minimum curvature (C_L) makes it possible to get the maximum diameter (D_L), which does not correspond to the egg length, as eggs are not generally spherical. The egg width is the place where the values of D_W and D_L are the highest (points in the right anterior portion of a graph D_W-D_L) (Figure 1.9), which represents the flattest part of the egg in all the spatial directions and is thus the middle of the egg.

On the basis that an egg has a circular transversal section, the value of D_W corresponds to the egg width. To estimate the egg length, we cannot use the same method, because it is only applicable to a circular section, and as eggs are not spherical, we cannot consider that the longitudinal section is circular. We then use the 1.48 ratio proposed by Williams (1981) to calculate the length (L) from the average width value obtained from a great number of egg shells. This ratio was established from extant large terrestrial birds (emus, cassowaries). Thus, from fossil egg fragments, we get the egg size corresponding to the length and width.

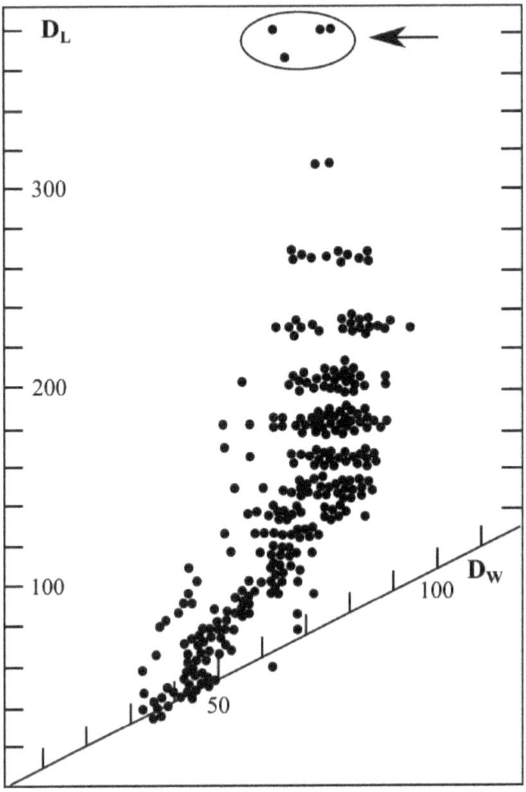

Figure 1.9. *Estimation of the width of an egg by means of the graph representing the transversal diameter (D_W) relative to the longitudinal diameter (D_l) for extant emus (according to Williams, 1981). The arrow corresponds to the flattest shell fragments for which the values of D_W will provide the egg width*

1.6.1.4.2. Volume of fossil eggs

In order to estimate the volume of an egg, the equation [1.8] proposed by Hoyt (1979) can be used, where K_v is a factor determined by Hoyt and equal to 0.51, L is the egg length and B the egg width, both measured in centimeters.

$$V_{egg} = K_v \times L \times B^2 \tag{1.8}$$

1.6.1.4.3. Mass of fossil eggs

Egg mass can be estimated by means of two different equations. The first one, proposed by Hoyt (1979) [1.9], directly links the egg mass to the egg size, with K_w which is a factor determined by Hoyt and equal to 0.55, L the length and B the width of the egg measured in centimeters.

$$M_{egg} = K_w \times L \times B^2 \quad [1.9]$$

The second equation [1.10] allowing us to estimate the egg mass is proposed by Dyke and Kaiser (2010), and links the egg mass (M_{egg}) to the femur length (L_{fem}) of the bird that laid it (Figure 1.10). Thus, by knowing the femur length of the female, it is possible to estimate the mass of the eggs it laid, and also by knowing the fossil egg mass, it is possible to estimate the femur length of the female that laid it.

$$\log(L_{egg}) = 0.4503 \times \log(M_{egg}) + 1.0336 \quad [1.10]$$

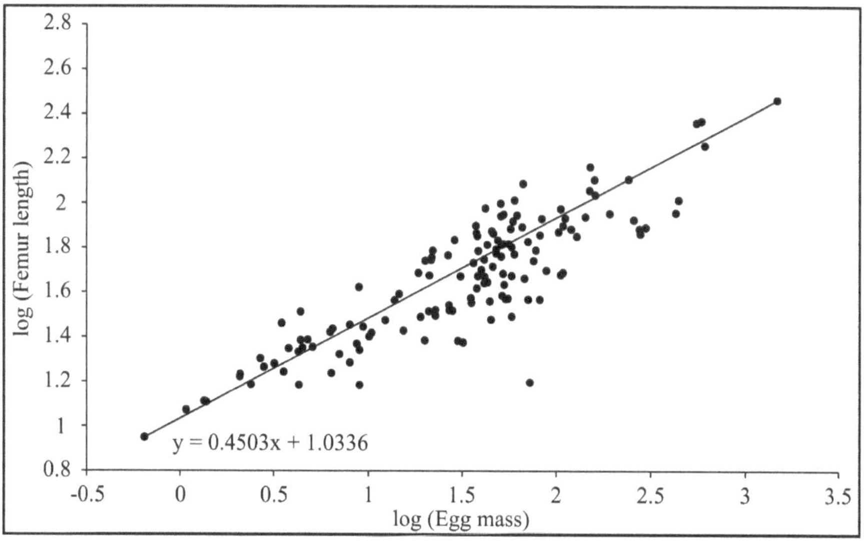

Figure 1.10. *Estimation of the mass of an egg from the femur length of the female (according to Dyke & Kaiser, 2010)*

1.6.1.4.4. Incubation duration

The incubation period of a fossil bird egg can be estimated by means of various very similar equations, the most recent and robust one (with an R^2 of 0.68) being that proposed by Rahn and Ar (1974). This equation, established from the study of extant birds, links the egg weight (W, in grams) to the incubation duration (I, in number of days) necessary for this bird embryo to be fully formed (equation [1.11]).

$$I = 12.03 \times W^{0.217} \quad [1.11]$$

1.6.2. *Isotope geochemistry*

Isotope geochemistry is a tool enabling us to reconstruct the paleobiology of birds and the paleoenvironment in which they lived, using the proportions of heavy stable isotopes compared with light stable isotopes present in the fossil bone tissues or in the eggshells (e.g. the heavy stable isotope of carbon (^{13}C) compared with the light stable isotope of carbon (^{12}C), thus written $^{13}C/^{12}C$). These values are written in delta unit (δ) and expressed in per thousand (‰). Between each reservoir, this proportion changes according to physicochemical conditions, which correspond to isotope fractionation. By knowing the value of this fractionation between two reservoirs (Δ) and by knowing the δ value of one of the reservoirs, it is then possible to determine the δ value of the other reservoir (Figure 1.11).

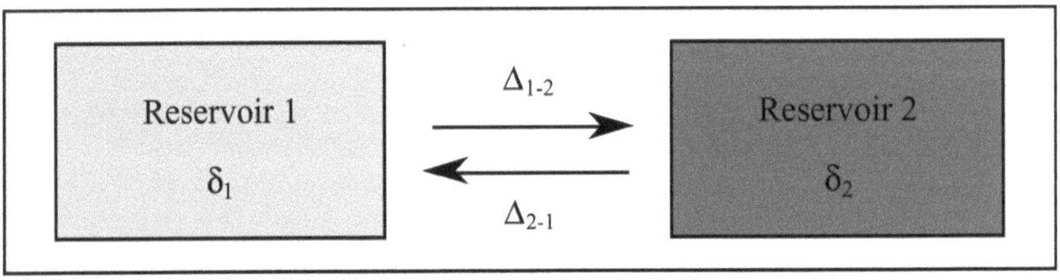

Figure 1.11. *Diagram illustrating the isotope fractionation between two reservoirs. For a color version of the figure, see www.iste.co.uk/angst/birds.zip*

For all isotope analyses, a sample of a small quantity of bone powder, tooth enamel, dentine or eggshell is necessary. After chemical pretreatments, specific to the study of each chemical element, part of this powder is analyzed by mass spectrometry.

1.6.2.1. *$\delta^{13}C$ and diet estimation*

The reconstruction of the diet of fossil birds is possible through the study of the carbon isotopes present in the hydroxyapatite of fossil bones or in the carbonates of fossil eggshells. The carbon isotope composition is directly linked to diet enriched in ^{13}C (compared to ^{12}C) between apatite and diet (Koch, 2007). The value of this enrichment was measured for extant birds with known diets and for which $\delta^{13}C$ measurements were taken from bones or eggshells, and from their food. Thus, in the case of herbivorous birds, the enrichment between mineralized tissues (bone or shell) and food ($\Delta_{\text{mineralized tissue-food}}$) ranges from –13.4‰ to –16.2‰ (i.e. an average value of –14.8‰), and it is of –8.3‰ for carnivorous birds (Angst *et al.*, 2014; Figure 1.12). For mammals, the same studies were carried out and made it possible to estimate that the carbon isotope fractionation ($\Delta_{\text{mineralized tissue-food}}$) of herbivorous mammals ranges from –11.4‰ to –14.6‰ (i.e. an average value of –13‰) (Passey *et al.*, 2005), and a value of –9‰ for carnivorous mammals (Lee-Thorp *et al.*, 1989). From the $\delta^{13}C$ value of the fossil mineralized tissue and using isotope fractionation, we can then calculate the $\delta^{13}C$ value of the food consumed by these animals. The value obtained for birds can then be compared with the values obtained for other animals with a known diet and present at the same fossil sites, such as mammals (herbivores or carnivores, determinable from the study of teeth), crocodiles (carnivores), etc., and also with the range of values known in the present for plants, which ranges from –35‰ to –9‰. Thus, if the $\delta^{13}C$ value of food is consistent with the $\delta^{13}C$ values of the known plants at that time, but also with the $\delta^{13}C$ value of food obtained from herbivorous mammals from the same site, it can be concluded that the bird studied had a herbivorous diet. Conversely, if this value is consistent with values obtained for carnivorous mammals from the same site, we can consider that this bird was carnivorous.

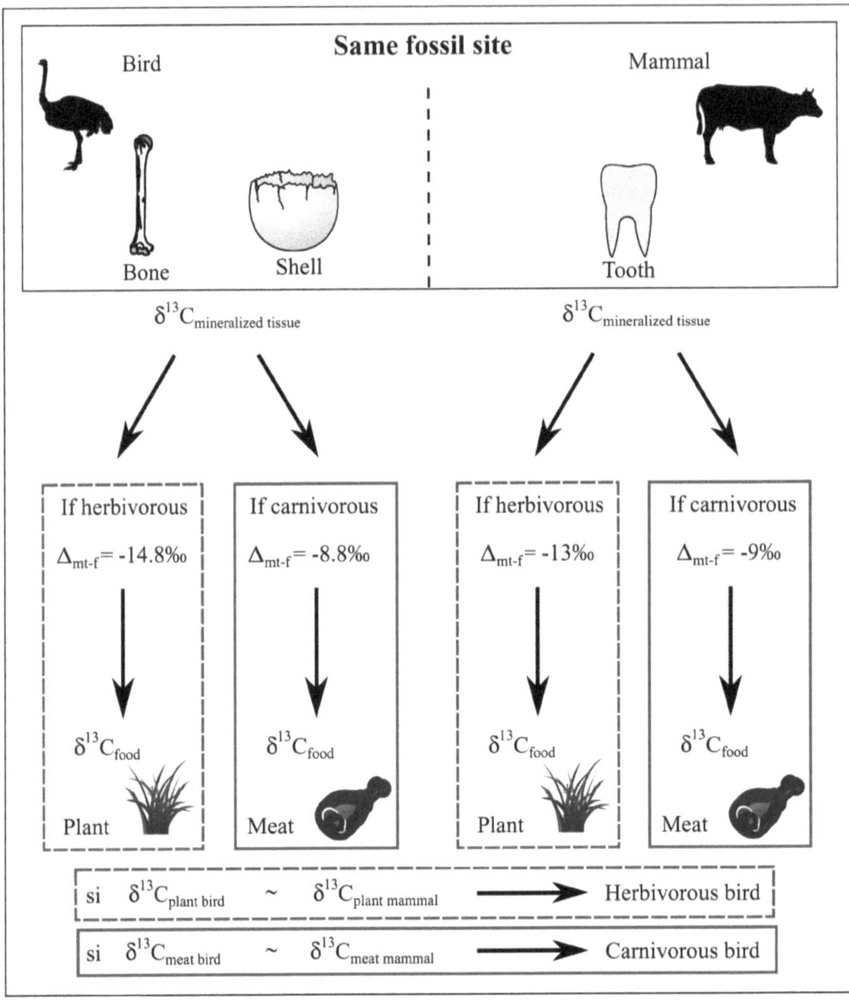

Figure 1.12. *Diagram of the estimation of a bird diet from the $\delta^{13}C$ of its mineralized tissues (mt) (eggshell or bone) compared with the mineralized tissues of mammals (tooth) from the same fossil site. For a color version of the figure, see www.iste.co.uk/angst/birds.zip*

In the case of herbivorous birds, if we consider that the animal consumed all the plants in its environment without any food preferences, the $\delta^{13}C$ value of plants (estimated from the fossil mineralized tissues) will also allow us to estimate the proportion of C3 or C4 plants present in the environment, as these two types of plants do not have the same values of $\delta^{13}C$ ($\delta^{13}C_{C3}$

plants ~ -26‰, $\delta^{13}C_{C4}$ plants ~ -13‰; O'Leary, 1988; Kennedy & Krouse, 1990; Cerling *et al.*, 1997; Cerling & Harris, 1999).

Conversely, if we consider that these birds preferred consuming a certain type of plant when compared with all the plant species available in their living environment, it is also possible to estimate the proportions of each type of plants consumed, using the following system of equations (equations [1.12] and [1.13]) (Clarke *et al.*, 2006):

$$\delta_M = F_A\delta_A + F_B\delta_B \quad [1.12]$$

$$1 = F_A + F_B \quad [1.13]$$

with δ_M corresponding to the $\delta^{13}C$ value of the mix of plants consumed by the bird, that is, the value of $\delta^{13}C$ measured from the mineralized tissues of this bird (in ‰), F_A corresponding to the flow (in % of contribution) of the first source available and F_B corresponding to the flow (in % of contribution) of the second source available in the environment (which can be C3, C4 or CAM plants) and for which the values of $\delta^{13}C$ are measured in the environment. Thus, the proportions of type A and type B plants consumed can be calculated using equations [1.14] and [1.15]:

$$F_A = 1 - \left(\frac{\delta_M - \delta_A}{\delta_B - \delta_A}\right) \quad [1.14]$$

$$F_B = \frac{\delta_M - \delta_A}{\delta_B - \delta_A} \quad [1.15]$$

1.6.2.2. *$\delta^{15}N$ and trophic level estimation*

Diet or more specifically trophic level can also be estimated thanks to nitrogen isotopes ($\delta^{15}N$) derived from amino acids composing proteins. Using nitrogen isotopes requires then a preservation of these amino acids, which is only possible for recent fossils, kept in suitable environmental conditions. In cases where these isotopes are preserved, there is a known isotope fractionation of 3‰ between food and tissues due to a biochemical fractionation during the production of tissues (Koch *et al.*, 1994; Johnson *et al.*, 1997) (Figure 1.13). Thus, as we know the $\delta^{15}N$ value of fossil mineralized tissues and using this isotope fractionation, we can find the $\delta^{15}N$ of the food consumed by this animal. It is then possible to know whether the animal is a super-predator at the top of the food chain or a herbivore at the bottom of the food chain, according to the more or less high value of $\delta^{15}N$ in its tissues.

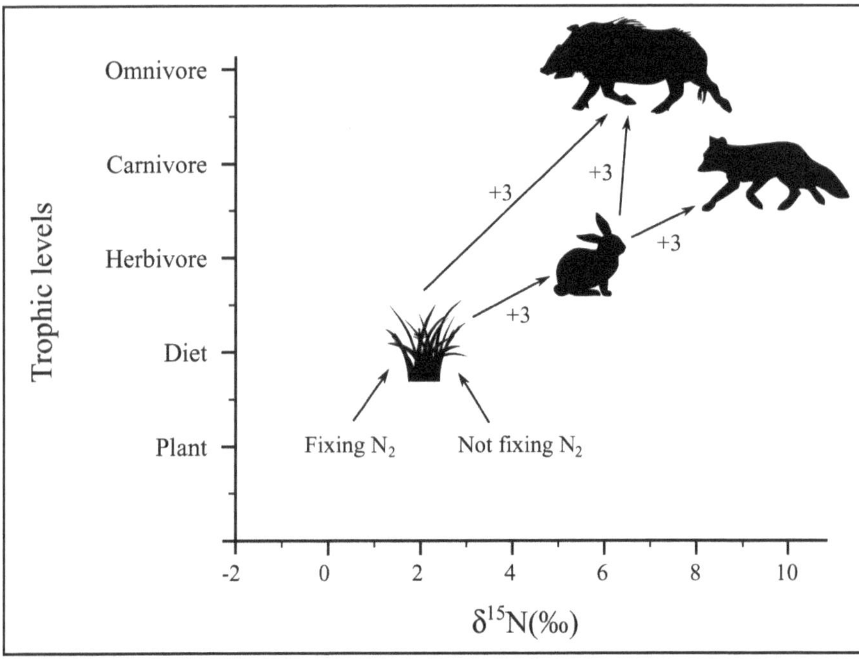

Figure 1.13. *Evolution of $\delta^{15}N$ and isotope fractionation according to different trophic levels (according to Koch* et al., *1994)*

1.6.2.3. *$\delta^{13}C$ and estimation of paleoprecipitations*

In the case of herbivorous birds, as discussed previously, the $\delta^{13}C$ value of mineralized tissues (bones or eggshells) allows us to calculate the $\delta^{13}C$ value of the food consumed, which are then the plants in the environment, if we consider that the animal consumes all the plants in its environment, with no food preferences. Diefendort *et al.* (2010) and Kohn (2010) established a correlation between the $\delta^{13}C$ value of plants and mean annual precipitations (MAP; expressed in millimeters per year), which was modified by Rey *et al.* (2013) (equation [1.16], Figure 1.14).

$$log_{10}(MAP + 300\ (mm.y^{-1})) = 0.092(\pm 0.004)\Delta^{13}C_{leaves} + 1.148(\pm 0.074) \quad [1.16]$$

with $\Delta^{13}C_{leaves} = (\delta^{13}C_{atm} - \delta^{13}C_{leaves}) / (1 + \frac{\delta^{13}C_{leaves}}{1000})$. The value of $\delta^{13}C_{leaves}$ corresponds to the $\delta^{13}C$ value of plants consumed by the bird estimated using the method described previously. The $\delta^{13}C_{atm}$ value corresponds to the value of atmospheric $\delta^{13}C$, which can be estimated through different independent studies. Thus, by knowing the $\delta^{13}C$ value of the plants in the living environment of the bird studied, it is possible to know the mean annual precipitation of this environment.

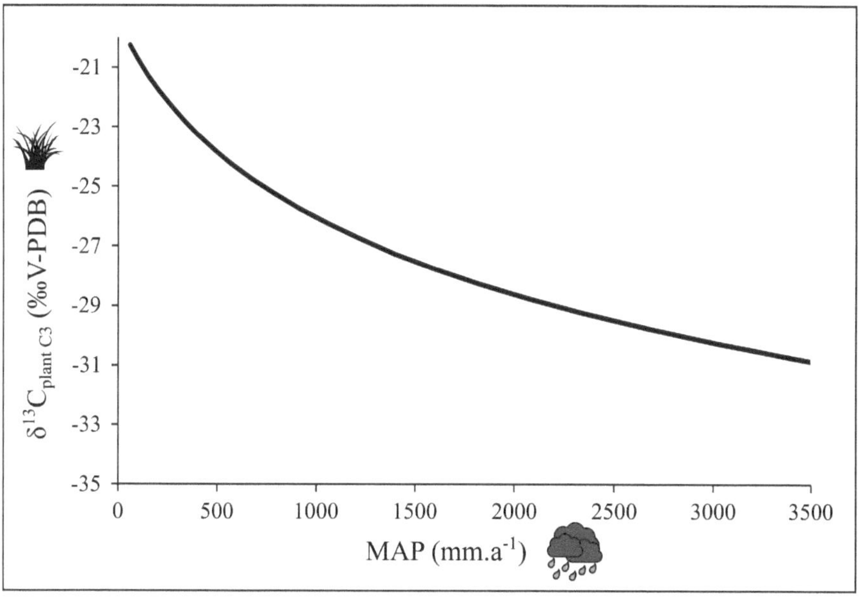

Figure 1.14. *Graph of the $\delta^{13}C$ of the C3 plants as a function of the mean annual precipitations (MAP) according to the equation formulated by Rey* et al. *(2013). For a color version of the figure, see www.iste.co.uk/angst/birds.zip*

1.6.2.4. *$\delta^{18}O$ and estimation of paleotemperatures*

The estimation of paleotemperatures can be made thanks to the $\delta^{18}O$ measurement of the mineral phase (carbonate) of fossil eggshells. Eggshells consist of 95% of calcium carbonate, the oxygen of which is directly derived from body water, which itself is derived from the water drunk by the bird just before and at the time of the egg formation (Von Schirnding *et al.*, 1982). We consider that the water consumed by birds mainly corresponds to rainwater, also called meteoric water (mw). Between each of these stages, an isotope fractionation occurs (Figure 1.15). The values of these fractionations can vary according to a large number of physiological factors (e.g. the size of the birds, or even their more or less significant adaptation to an aquatic environment (Lazzerini *et al.*, 2016). In order to limit the errors linked in particular to the size of birds, the isotope fractionations used in the context of paleotemperature estimation from large fossil birds were calculated from the study of extant ostriches (Angst *et al.*, 2015). A first isotope fractionation between body water (bw) and mineralized tissues (mt; eggshell) ($\Delta^{18}O_{mt\text{-}bw}$) was calculated by means of the $\delta^{18}O$ measurements of the eggshell of numerous extant ostriches associated with the body water $\delta^{18}O$ of these same birds, measured from the analysis of the $\delta^{18}O$ of the water extracted from the white and yolk of the eggs. $\Delta^{18}O_{mt\text{-}bw}$ has a value of +30‰ (Angst *et al.*, 2015). Thus, the $\delta^{18}O$ value of the bird body water can be calculated from the $\delta^{18}O$ value of eggshells ($\delta^{18}O_{mt}$) according to equation [1.17]. It is important to note that $\delta^{18}O_{bw}$ is reported relative to an international standard, different from the eggshell $\delta^{18}O$ ($\delta^{18}O_{mt}$). Indeed, depending on conventions, the former is expressed in relation with SMOW (V-SMOW), whereas the latter is expressed in relation with PDB (V-PDB). Before using equation [1.17], we must then transform the value of $\delta^{18}O_{mt}$, according to equation [1.18] proposed by Coplen *et al.* (1983).

$$^{18}O_{bw} = {}^{18}O_{mt} - \Delta^{18}O_{mt-bw} \qquad [1.17]$$

$$\delta^{18}O(\text{‰V} - \text{SMOW}) = \delta^{18}O\ (\text{‰V} - \text{PDB}) \times 1.03091 + 30.91 \qquad [1.18]$$

The second isotope fractionation, between the water ingested by birds and body water ($\Delta^{18}O_{bw\text{-}mw}$), was also estimated by knowing the $\delta^{18}O$ value of the rainwater drunk by birds for which the previous measurements were taken. $\Delta^{18}O_{bw\text{-}mw}$ is then worth + 4‰ (Angst *et al.*, 2015). Thus, the $\delta^{18}O$ value of the meteoric water that the bird ingested can be calculated from the $\delta^{18}O$ value of the bird body water according to equation [1.19].

$$\delta^{18}O_{mw} = \delta^{18}O_{mw} - \Delta^{18}O_{bw-mw} \qquad [1.19]$$

Once the $\delta^{18}O$ of meteoric water is calculated from the isotope fractionations established in the present, we can estimate the paleotemperature of the environment in which these birds laid eggs. Actually, there are two connections allowing us to link the $\delta^{18}O$ values of meteoric water to the temperature of the environment. Lécuyer (2013) proposed equation [1.20] making it possible to link the $\delta^{18}O_{mw}$ in ‰ to the mean annual temperature (MAT). The second equation [1.21] allows us to estimate the mean summer temperature (MST) (i.e. the hottest months) from the $\delta^{18}O_{mw}$ in ‰ (Bernard *et al.*, 2009) (Figure 1.16).

$$MAT\left(°C\right) = \left({}^{18}O_{mw} + 14.172\right) / 0.562 \quad 1.20]$$

$$MST\left(°C\right) = 1.06 \times {}^{18}O_{mw} + 24.1 \quad [1.21]$$

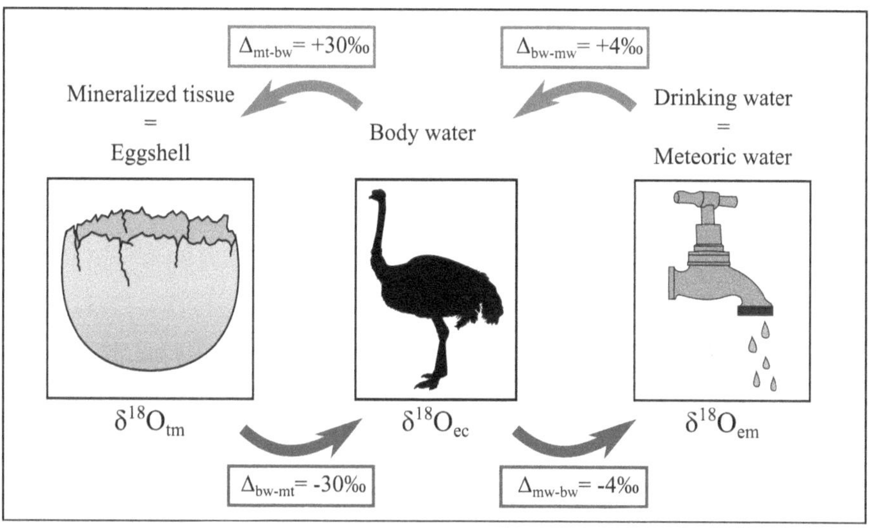

Figure 1.15. *Diagram of the different estimation stages of the $\delta^{18}O$ of meteoric water from the $\delta^{18}O$ of eggshells. mt: mineralized tissue; bw: body water; mw: meteoric water; green arrows: isotope fractionation values established from extant ostriches; red arrows: isotope fractionation values to be used to estimate the $\delta^{18}O$ of meteoric water from the $\delta^{18}O$ of fossil eggshells. For a color version of the figure, see www. iste.co.uk/angst/birds.zip*

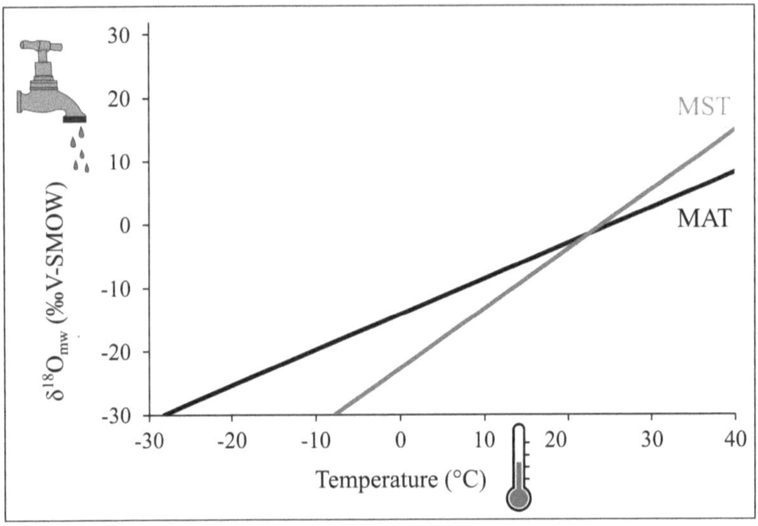

Figure 1.16. *Graph of the $\delta^{18}O$ meteoric water (mw) compared with the mean annual temperature (MAT, black line) according to Lécuyer (2013), or compared with the mean summer temperature (MST, gray line) according to Bernard* et al. *(2009). For a color version of the figure, see www.iste.co.uk/angst/birds.zip*

1.6.3. *Functional morphology*

1.6.3.1. *Study of musculature and diet*

Part of functional morphology corresponds to the study of the skeletal musculature, and the link between these muscles and the movements they facilitate. Thus, the study of the muscles of the skull and the mandible allows us to estimate the diet, in particular the adductor muscle of the mandible, which allows the jaws to close, and the pterygoid muscle, which allows birds to move the beak tip. Among extant granivorous birds, the adductor muscle of the mandible is very developed compared with the pterygoid muscle (Figure 1.17(B)). This is due to the fact that granivorous birds need a high grinding capacity to crush the grains making up their diet, but do not necessarily need to use the tip of their beak to collect them. Conversely, among extant carnivorous birds, the adductor muscle of the mandible is small in size and the pterygoid muscle is very developed (Figure 1.17(A)). Carnivorous birds use the tip of their beak to quickly kill their prey and to rip pieces of flesh. However, they do not chew their food, but swallow it whole and do not therefore need a significant crushing power. It is sometimes possible to see traces left by these muscles on bones, on very well-preserved skulls and mandibles of fossil birds (Figure 1.17(C)). According to the size of the muscular insertions, it is possible to estimate the relative size of these two muscles, and thus, the diet of fossil birds.

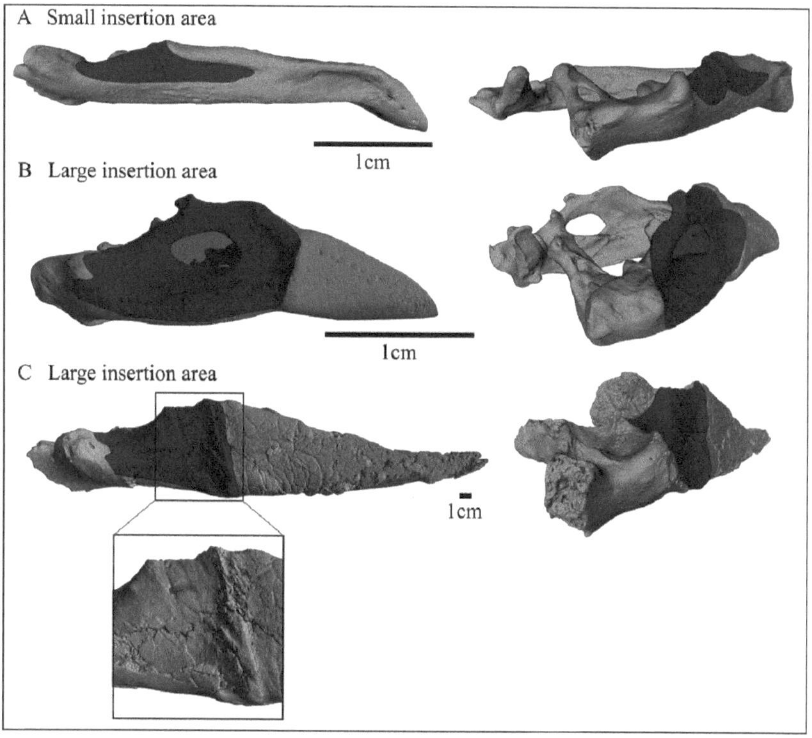

Figure 1.17. *Comparison between the insertion area of the adductor muscle of the mandible (dark area) for extant birds and* Gastornis *in lateral view (left part) and in oblique view (right part): (A) Extant carnivorous bird (*Buteo platypterus*), (B) extant granivorous passerine (*Geospiza magnirostris*) and (C)* Gastornis *mandible and zoom on the muscular insertion area allowing us to estimate the size of the adductor muscle of the mandible of this fossil bird (according to Angst* et al.*, 2014)*

1.6.3.2. *Finite element analysis (FEA)*

Finite element analysis (FEA) is a method derived from engineering, which has been adapted to the study of fossils. The goal of this method is to study the physical reaction of an object when it is subjected to an external force. In the case of engineering, the goal was to test, for example, the resistance of a bridge to the weight of cars crossing it. In the case of fossils, this approach was modified to test the physical resistance of the skeletal elements when they are subjected to different external forces, for example, the resistance of leg bones compared to the weight of the animal, or the biting capacity of the jaw according to food hardness. In order to carry out this method, a 3D model of the study object is required. As only the surface of the object will be used, a surface scan is sufficient (Figure 1.18(1)). This surface 3D model is then subdivided into a large number of polygons called "Finite Elements", which form the surface mesh of the fossil. Each of these finite elements is composed of nodes corresponding to the vertices of polygons and sides corresponding to the connections between nodes (Figure 1.18(2)). The fossil will then be modeled as a surface composed of a set of small polygons. The next stage involves "filling" this surface form with a predetermined number of identical polygons in order to obtain a volume mesh and hence a complete 3D model of the specimen (Figure 1.18(3)). We must then characterize the model by specifying the physical characteristics of the bone, such as the modulus of elasticity, Poisson's ratio or density, which are estimated from the studies of extant birds (Figure 1.18(4)). The model itself is then ready and initial conditions, also called the boundary conditions of the model, should be determined. This corresponds to the definition of the areas where the bone in question is linked to another bone by an articulation, and also to muscular insertion areas. We must then determine the forces that will be applied to this model, that is to say their point of application, direction and intensity (Figure 1.18(5)). Once the simulation is launched, the force will cause a movement of the finite elements, which will be directly linked to the force, and the whole model will undergo a deformation, as all the model meshes are linked. The study of the movement of each mesh will make it possible to estimate the overall deformation, stress and pressure on this object. These factors will then be studied by means of the variation obtained for each node coordinate and will be viewable on a deformation, stress or pressure map of the specimen (Figure 1.18(6)). On these maps, colors will correspond to the intensity of the experienced deformation, stress or pressure. Most of the time, blue will indicate zero deformation, up to red or even white, which will indicate a maximum deformation. Therefore, this approach makes it possible to increase the intensity of the force up to the virtual breaking of the bone, and thus, test the resistance limits of the fossil.

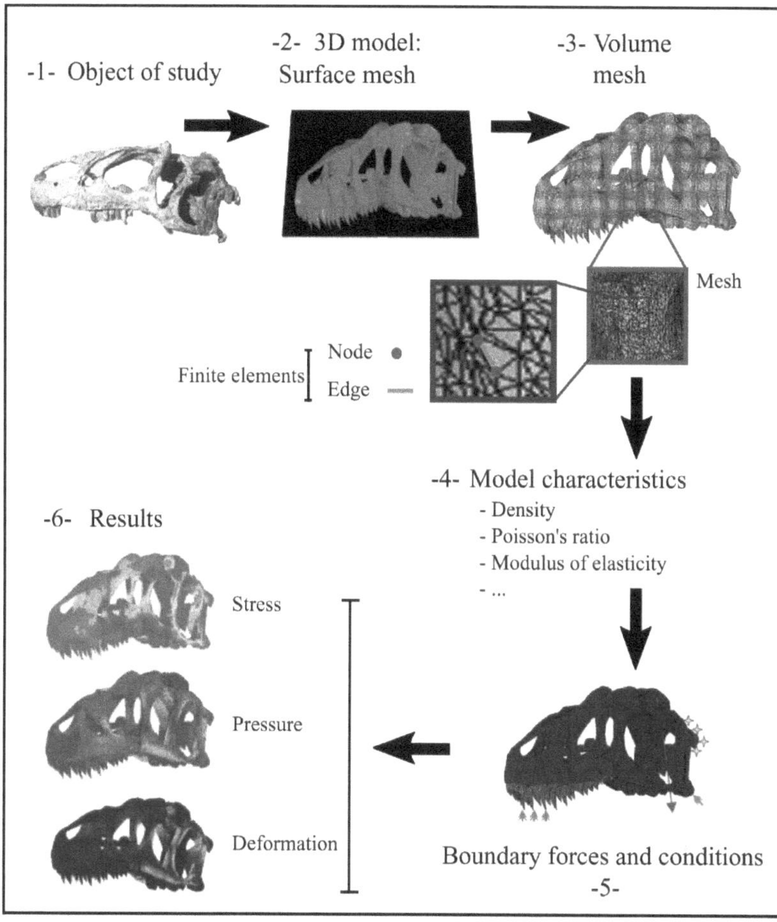

Figure 1.18. *Successive stages of the finite element analysis (after Rayfield, 2007). For a color version of the figure, see www.iste.co.uk/angst/birds.zip*

1.6.4. *Bone histology*

Bone histology is the study of bone tissue at different scales using optical tools. It can be subdivided into two parts: the study of bone microanatomy and the histological study itself. Microanatomy corresponds to the study of the global structure of the bone, such as the thickness of the bone wall, the shape of the bone section, the bone compactness or the presence of structures inside the bone cavity (medullary cavity). On the other hand, the histological study *sensu stricto* corresponds to the study of the bone tissue organization, focusing in particular on the shape, distribution, direction or density of bone cells, collagen fibers, and to the study of the depositional speed of the bone or the type of bone tissue present.

All these studies are based on the study of thin sections. These sections are made from an extant or fossil bone sample. This sample can be a complete bone section or a core drilled through the bone all the way to the medullary cavity (inside the bone) while preserving the rest of the bone intact. This second process is generally used on bones large enough to bear drilling, and too precious or rare to cut a complete portion of the bone. Once the sample is taken, if it concerns extant or subfossil bones, several ethanol and acetone baths must be carried out in order to remove the organic part of the bone. The samples are then embedded in a resin block in order to preserve their integrity during the following stages. Resin blocks are then cut, polished and glued on a glass slide. The sample is then thinned until a thickness allowing us to view the desired structures is obtained. The thin sections thus obtained can then be studied using an optical microscope.

Bone histology can provide various information on the paleobiology of fossil birds, as bone structures are well preserved in the fossil record. However, it was shown that, among all the vertebrates, including birds, bone structure can change in the same individual according to the bone studied (Chinsamy, 1995) or the sampling position on the same bone. This is why it is accepted that, for the paleoecological study of birds, hindlimb bones (femur, tibiotarsus and tarsometatarsus) are mainly used and that comparison between individuals or between species must be made between the same bones. In addition, by convention, samples are always taken from the middle of the bone diaphysis, so that they can be repeated and in order to get a proper comparison between several bones.

1.6.4.1. *Bone growth rates*

Bone histology can provide information on bone depositional rate. According to the speed of formation of a bone, its structure will not be the same. If the bone is formed very slowly, it will be composed of a bone with parallel fibers, or lamellar bone, in which collagen fibers are organized parallel to each other, with a marked organization of the osteocytes (cells composing and forming the bone tissue) (Figure 1.19(A)). If the formation speed is faster, the bone is composed of a fibrolamellar bone, which will not have an organization of collagen fibers, nor osteocytes (Figure 1.19(B)).

Knowing the growth speed of a bone can then allow us to estimate the relative speed at which a bird reaches a size relatively close to the adult size.

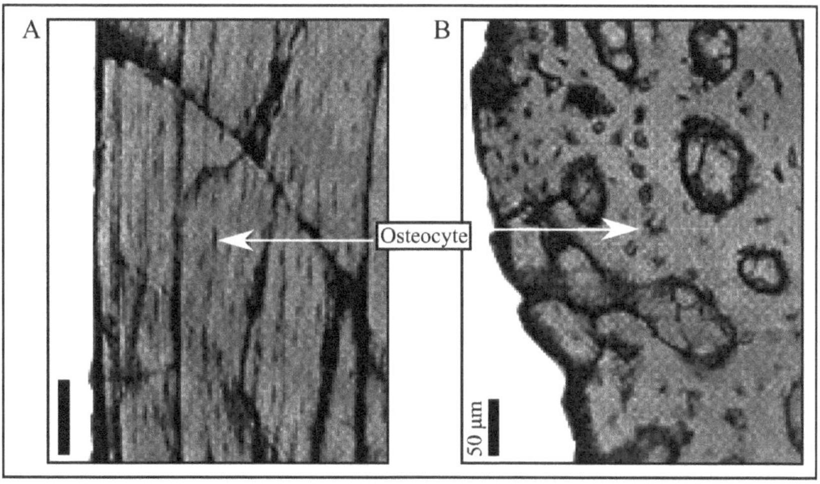

Figure 1.19. *Organization of the two main types of tissues observed among birds: (A) lamellar bone in the tibia of a* Gobipteryx *embryo with organized osteocytes and (B) fibrolamellar bone in the femur of an enantiornithine with osteocytes but no specific organization*

1.6.4.2. *Age and sexual maturity of individuals*

Thanks to the study of the bone structure of fossil birds, we can also get information on the age of the individuals and their sexual maturity degree (Chinsamy, 1995). Indeed, among the majority of birds, juvenile individuals go through, most of the time, first a fast bone growth stage composed of fibrolamellar bone, presenting a large number of primary osteons (corresponding to the primary vascularization of the bone, thanks to which nutrients are brought to the bone) (Figure 1.20(A)). During the growth of individuals, this primary tissue is going to be remobilized through resorption, when the rest of the organism requires calcium. This remobilization will take place, on the one hand, through the development of a secondary vascularization composed of secondary osteons, which are going to overlap the primary vascularization, and also through a medullary resorption of the bone, that is to say in the internal (also called endosteal) part of the bone. A secondary deposit of lamellar bone will then centripetally occur in the medullary cavity of the bone (central cavity). This deposit of endosteal lamellar bone then forms what is called the ICL (inner circumferential layer) (Figure 1.20(A)). When sexual maturity is reached, the external structure of the bone changes, from the remobilized primary fibrolamellar bone to a bone layer with a slower lamellar growth called the OCL (outer circumferential layer). The presence of any change in bone type between the fibrolamellar bone and the lamellar bone of the OCL then allows us to determine whether an individual has reached sexual maturity (Figure 1.20(B)).

After puberty, the bone will continue to grow widthwise with a slow deposition rate, which is going to gradually thicken the OCL. In certain cases, after sexual maturity, we see the formation of lines of arrested growth (LAGs) in the OCL, or sometimes in the deepest part of the cortex. These LAGs correspond to a slower bone growth up to a complete stop, and are most often interpreted as a reaction of the bird to poor annual environmental conditions (Castanet *et al.*, 1993). Thus, by counting the number of LAGs, it is possible to estimate the minimum age of the individuals, by considering that, for most birds, sexual maturity is reached during the first year after the hatching of eggs, and that birds form one line of arrested growth per year (Horner & Padian, 2004) (Figure 1.20(B)). We can only estimate the minimum age as, in some cases, some of the LAGs are erased by the remodeling of the bone. It is then difficult to be sure that we are actually seeing all the LAGs initially formed.

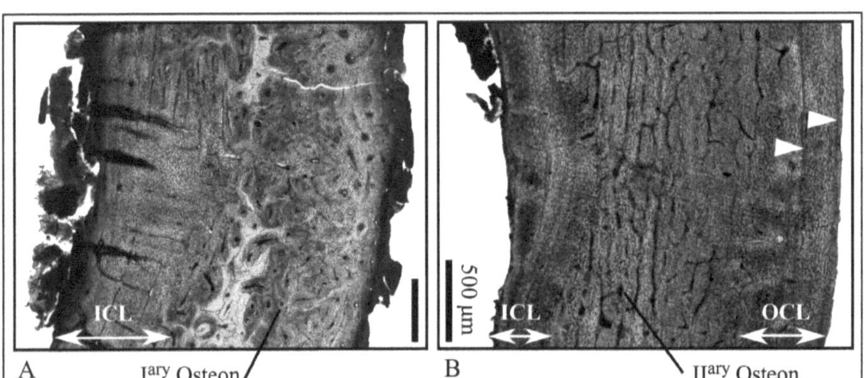

Figure 1.20. *Photograph of thin sections of* Raphus cucullatus. *A – Tibiotarsus of a juvenile. B – Femur of an adult showing lines of arrested growth in the OCL. ICL: inner circumferential layer; OCL: outer circumferential layer; white arrows: lines of arrested growth (LAGs); I^{ary} osteon: primary osteon; II^{ary} osteon: secondary osteon (from Angst* et al. *2017). For a color version of the figure, see www.iste.co.uk/ angst/birds.zip*

1.6.4.3. *Sex determination*

Another information that can be obtained thanks to bone histology is the sex determination of fossil individuals (Chinsamy *et al.*, 2013; Handley *et al.*, 2016). Among birds, when the female ovulates at the time of reproduction, it produces an additional quantity of bone, which will be used as a calcium reservoir in order to form the shell of its eggs. This additional supply of bone corresponds to the formation of a specific type of bone in the medullary cavity (central cavity of the bone), which is called medullary bone. This bone is of the fibrolamellar type, with no specific organization, and is an extension of the ICL. If medullary bone is identified in a specimen, it is then possible to assert that it is an ovulating female (Figure 1.21).

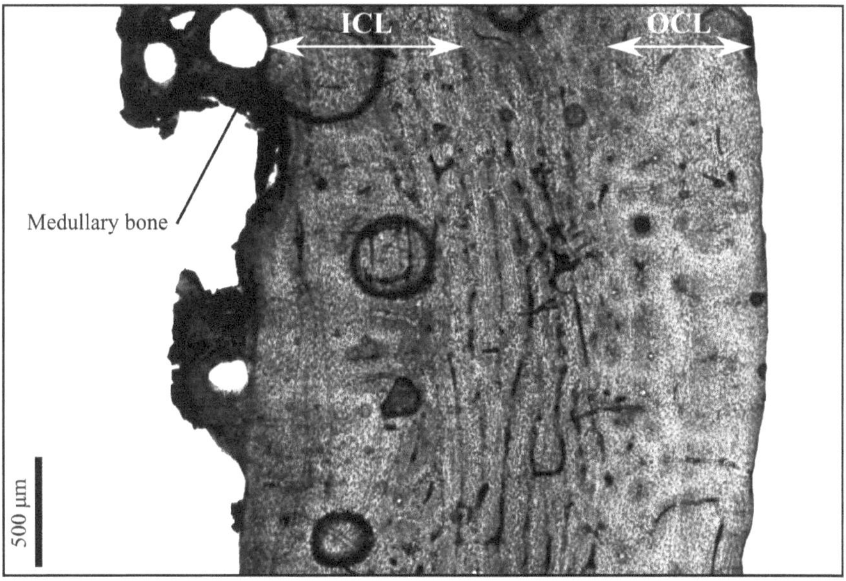

Figure 1.21. *Photograph of a thin section of a* Raphus cucullatus *femur with medullary bone indicating that it is an ovulating female (from Angst* et al. *2017). For a color version of the figure, see www.iste.co.uk/angst/birds.zip*

1.6.5. *General anatomy of birds*

This part is not intended to provide a list and give an exhaustive description of bird osteology, but only to illustrate the various anatomical terms that will be used in the following parts of this book (Figure 1.22). All the terminology used is derived from Baumel *et al.* (1993).

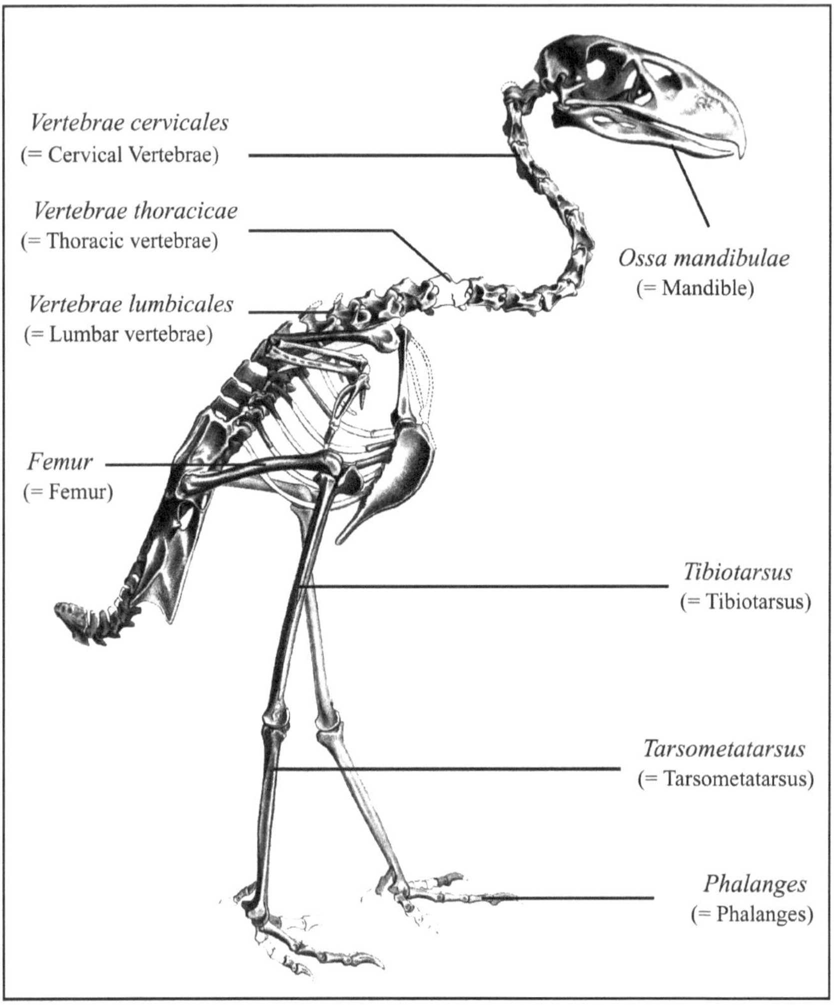

Figure 1.22. *Main bones used within the framework of this book, captioned on the skeleton of a* Psilopterus lemoinei *(Phorusrhacidae) derived from Sinclair and Farr (1932)*

1.7. Organization of the book

The following parts of this book will present the different groups of fossil and subfossil giant terrestrial birds in reverse chronological order, from the most recent to the oldest. We shall first discuss the Dinornithiformes, then the Aepyornithiformes, the Dromornithidae, the Phorusrhacidae, the

Brontornithidae, the Gastornithidae and finally *Gargantuavis*. For all these groups, a historical background is provided, followed by a summary of their systematics and their taxonomy, their stratigraphic and geographical distribution will be detailed, as well as their main anatomical characteristics. A summary of our knowledge on their biology and ecology will be presented, including in particular their diet, mode of locomotion, body mass and reproduction strategy. Finally, the various hypotheses regarding their extinction will be discussed.

2

Dinornithiformes

2.1. Historical background

The giant birds that once inhabited New Zealand came to the attention of Europeans when, during the 1830s, several travelers saw large bones in the hands of Maori people, and heard legends regarding giant birds such as the "moa" (Buick, 1931; Anderson, 1989; Berentson, 2012). It was in 1839 that Richard Owen published the first scientific description of such a bone, an incomplete femur provided by a certain Dr Rule, when he demonstrated at the end of a brilliant anatomical study that it belonged to a giant bird (Owen, 1839).

The discovery of moa bones increased during the 1840s, thanks in particular to the activities of William Mantell, who sent numerous specimens to England which had been collected in the rich deposits of the two main islands of New Zealand. Owen described these, as well as others subsequently collected, in a series of articles, gathered in two volumes in 1879 (Owen, 1879a). Other significant collections were sent to the natural history museums of other European countries, especially the museum in Vienna following the journey of exploration by Austrian geologist Ferdinand von Hochstetter. The 19th Century was also characterized by the development of moa studies carried out on site by New Zealand researchers, including Julius von Haast.

The scientific exploitation of a large number of fossil sites (swamps, caves and human habitation sites) facilitated the discovery, besides numerous skeletal remains, of mummified soft parts, feathers, eggs and footprints. It was also during the 19th Century that archeological studies on human populations that coexisted with moa and their interactions with these birds started.

This research continued throughout the 20th Century with large-scale excavations on sites such as the Pyramid Valley peatbog during the 1930s and 1940s, which improved understanding of the environments inhabited by moa. The development of radiometric dating methods, accompanied by archeological research, led to a much more accurate knowledge of the chronology of moa extinction caused by human predation. Since the beginning of the 21st Century, studies on fossil DNA have resulted in the clarification of these birds' systematics. Owing to the numerous modern investigation techniques applied to them, moa are certainly the best known of all extinct giant birds from a paleobiological point of view.

2.2. Taxonomy, systematics and classification

2.2.1. *Taxonomy*

The taxonomy and systematics of Dinornithiformes have been significantly clarified since studies on their DNA have been carried out (Huynen *et al.*, 2003; Allentoft & Rawlence, 2012). The authors of the 19th and 20th Centuries, relying on debatable differences in size and morphology, had multiplied taxa in a very exaggerated way (see Worthy and Holdaway (2002), for a history of moa taxonomy up to 2002). Thus, Oliver (1949) recognized within the order Dinornithiformes, 3 families, which themselves contained 7 genera, subdivided into 29 species! More recently, less typological classifications have been proposed using more sophisticated analysis techniques, taking into account individual variability and sexual dimorphism. Thus, Cracraft (1976) only recognized 13 species divided into 6 genera. Worthy and Holdaway (2002) recognized 6 genera and 11 species.

Paleogenetic research leads to the conclusion that there were, during the period covering the Late Pleistocene and the Holocene (for which we have usable fossil mitochondrial DNA), 9 species of Dinornithiformes belonging to 6 genera, namely *Dinornis*, *Megalapteryx*, *Pachyornis*, *Anomalopteryx*, *Emeus* and *Euryapteryx* (Bunce *et al.*, 2009). See the next section, on systematics, for details.

2.2.2. *Systematics*

After more than a century of different proposals from various authors, today moa systematics has reached a certain stability thanks to the study of ancient DNA, which makes it possible to synonymize a number of taxa previously considered as distinct. The systematics below follows that proposed by Worthy and Scofield (2012):

Class Aves Linnaeus, 1758

Super-order Palaeognathae Pycraft, 1900

Order Dinornithiformes Newton, 1884

Stratigraphic occurrence: Miocene to Holocene of New Zealand

Family Dinornithidae Bonaparte, 1849

Genus *Dinornis* Owen, 1843a

– *Dinornis novaezealandiae* Owen, 1843b

– *Dinornis robustus* Owen, 1846

Family Megalapterygidae Bunce *et al.*, 2009

Genus *Megalapteryx* Haast, 1886

– *Megalapteryx didinus* (Owen, 1883)

Family Emeidae Bonaparte, 1853

Genus *Emeus* Reichenbach, 1852

– *Emeus crassus* (Owen, 1846)

Genus *Euryapteryx* Haast, 1874

– *Euryapteryx curtus* (Owen, 1846)

Genus *Anomalopteryx* Reichenbach, 1852

– *Anomalopteryx didiformis* (Owen, 1844)

Genus *Pachyornis* Lydekker, 1891

– *Pachyornis geranoides* (Owen, 1848)

– *Pachyornis elephantopus* (Owen, 1856b)

– *Pachyornis australis* Oliver, 1949

2.2.3. *Classification*

As recognized by the very first studies on this matter (Owen, 1839), moa were related to extant ratites, that is, flightless paleognaths. It was often assumed (see, e.g. Bourdon *et al.* (2009)) that they were closely related to other endemic ratites from New Zealand, the kiwis (Apterygiformes), which might seem consistent from a biogeographical point of view. However, studies on the DNA of Dinornithiformes provided unexpected results (see Worthy & Scofield (2012) for a history of the research on this matter). In particular, the study by Phillips *et al.* (2010) on the mitochondrial genome came up with the tinamous, a group of South American paleognaths that are still able to fly, as the sister-group of moa, which implies that the loss of the ability to fly occurred several times within "ratites" (Figure 2.1). Mitchell *et al.* (2014), also using mitochondrial DNA, found the tinamous as the sister-group of moa too, and considered kiwis as the sister-group of Madagascan Aepyornithiformes (Figure 2.1). These results thus suggest complex paleobiogeographical histories for the New Zealand "ratites".

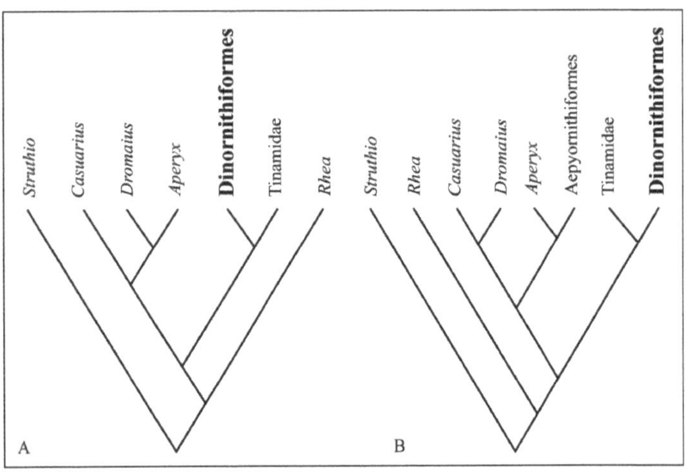

Figure 2.1. *Phylogenetic position of Dinornithiformes according to the hypotheses of Phillips* et al. *(2010, A), and of Mitchell* et al. *(2014, B)*

2.3. Stratigraphic and geographical distribution

Dinornithiformes are only known from the New Zealand archipelago (North Island, South Island, Stewart Island and some other small islands) (Figure 2.2). The existence of moa in Australia was assumed by De Vis

(1884), on the basis of an incomplete femur, supposedly from Queensland, which he called *Dinornis queenslandiae*. The specimen was the subject of various interpretations and some reckoned it to be a cassowary bone (Hutton, 1893). Scarlett (1969) convincingly interpreted it as most likely a bone of the moa *Pachyornis elephantopus*, broken by a human, coming from a New Zealand anthropic site, and mixed with Australian material.

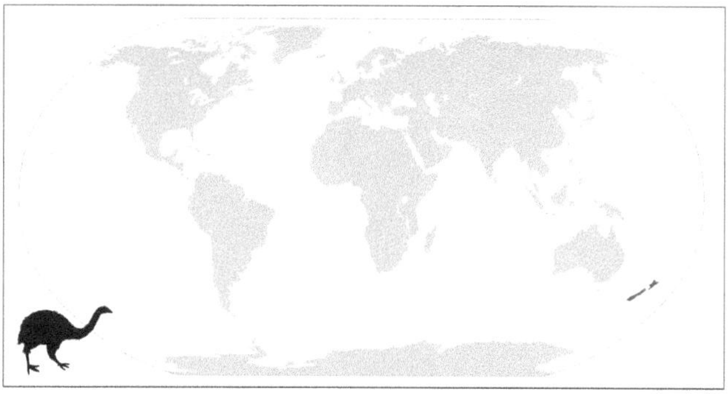

Figure 2.2. *Geographical distribution of Dinornithiformes. For a color version of the figure, see www.iste.co.uk/angst/birds.zip*

The vast majority of sites that provided moa remains are Quaternary (Pleistocene or Holocene) in age. The older, Miocene, age of the Saint Bathans deposits is exceptional.

2.3.1. *Miocene*

The oldest known remains of Dinornithiformes come from the Saint Bathans site, in the province of Otago, on the South Island of New Zealand (Figure 2.3). Remains of vertebrates come from sand and clay deposits belonging to the Bannockburn Formation, deposited near the shores of a large freshwater lake and dating from the end of the Early Miocene and the beginning of the Middle Miocene (–16 to –19 Ma) (Worthy *et al.*, 2007). The deposit yielded bird remains belonging to at least 23 species, including moa. The latter were represented by eggshell fragments and bone fragments, especially of tibiotarsi (Tennyson *et al.*, 2010). This material indicated the presence of two moa species, including one weighing approximately 20 to 25 kg, while the other was much larger and thus certainly unable to fly. This suggests that the evolutionary history of moa began a very long time ago, probably before the Late Oligocene, during which there was a significant but incomplete submersion of "Zealandia".

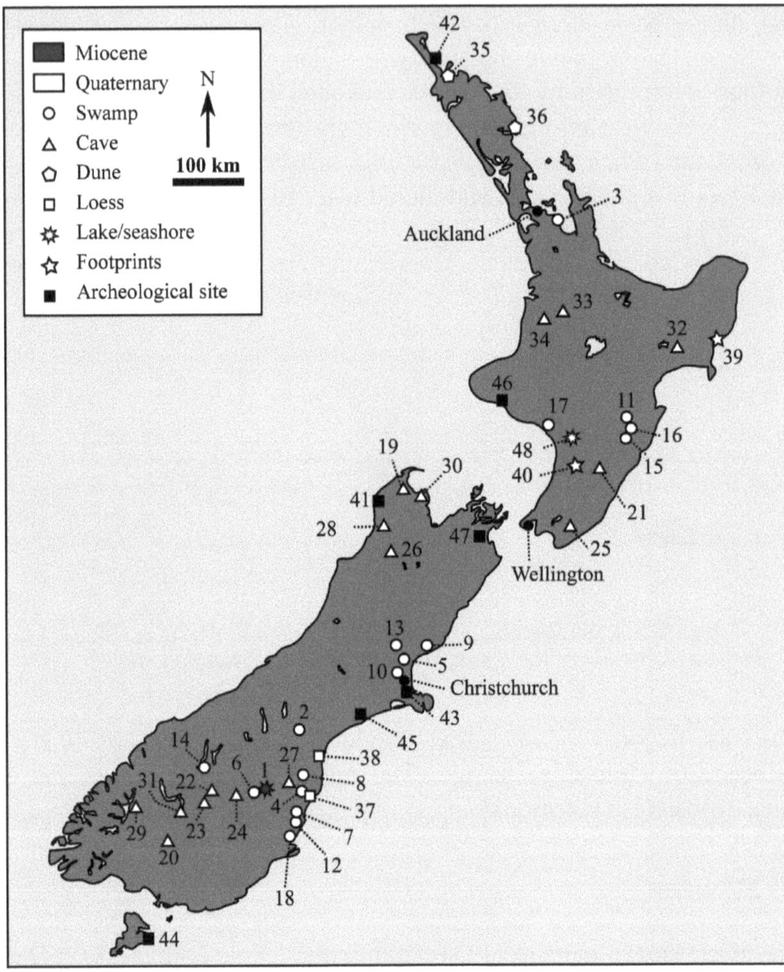

Figure 2.3. *Geographical distribution of the main sites where Dinornithiformes were discovered. Miocene lake shore site: 1 – Saint Bathans; Quaternary swamp sites: 2 – Albury Park, 3 – Clevedon, 4 – Enfield, 5 – Glenmark, 6 – Hamilton, 7 – Herbert, 8 – Kapua, 9 – Motunau, 10 – North Dean, 11 – Poukawa, 12 – Pukemata, 13 – Pyramid Valley, 14 – Scaife's Lagoon, 15 – Tangatupura, 16 – Te Aute, 17 – Upokongaro, 18 – Waikouaiti; Quaternary cave sites: 19 – Aorere, 20 – Castle Rock, 21 – Coonoor, 22 – Cromwell Gorge, 23 – Earnscleugh, 24 – Manuherikia, 25 – Martinborough, 26 – Mont Owen, 27 – Ngapara, 28 – Oparara, 29 – Takahe Valley, 30 – Tarakohe, 31 – Waikaia, 32 – Waikaremoana, 33 – Waipa, 34 – Waitomo; Quaternary dune sites: 35 – Doubtless Bay, 36 – Pataua; Quaternary loess sites: 37 – Cape Wanbrow, 38 – Timaru; Quaternary footprint sites: 39 – Gisborne, 40 – Palmerston North; Archeological sites: 41 – Heaphy, 42 – Houhora, 43 – Redcliffs, 44 – Ringaringa, 45 – Wakanui, 46 – Waingongoro, 47 – Wairau Bar; Quaternary seaside site: 48 – Marton*

2.3.2. *Early Pleistocene*

Pleistocene sites prior to the last glacial period (Otiran) are relatively few (33 according to Worthy *et al.* (1991)). Some date back to the Early Pleistocene and are approximately 2 million years old (Tewkesbury Formation). The Marton site, on the North Island (Figure 2.3), dating back approximately 1 million years, provided the remains of two moa taxa, including *Anomalopteryx didiformis*. It can be noted that all the bones identifiable from the Lower Pleistocene deposits can be assigned to species that were still present during the Holocene, in particular *Dinornis novaezealandiae*, *Euryapteryx geranoides*, *Euryapteryx curtus* and *Anomalopteryx didiformis*. Therefore, these species have long evolutionary histories.

2.3.3. *Late Pleistocene and Holocene*

The vast majority of the very abundant known moa remains come from Late Pleistocene (corresponding to the last glacial period) and Holocene sites, up to the extermination of this bird group by the Maoris, at an unclear date, but posterior to the 14th Century A.D. The deposits corresponding to these periods are extremely numerous, spreading over all regions of New Zealand (Figure 2.3), and providing large quantities of moa bones, corresponding to thousands of individual specimens (see Anderson's tables, 1989). Therefore, it is not possible to mention them all, and it seems more interesting to differentiate the main types of sites that provided moa remains, as Oliver (1949) and Anderson (1989) did. Some of these sites are considered as "natural", as moa died there of natural causes, while others reflect the hunting activities of the Maoris.

2.3.3.1. *Dunes and other surface sites*

When Europeans arrived, many moa bones were discovered lying on the ground surface, under a thin layer of plant remains, which had apparently remained in the very place where the birds had died. In particular, numerous skeletons buried in coastal dunes were discovered, often associated with egg fragments. In Canterbury province, moa remains were found in loess deposits (Oliver, 1949; Figure 2.3).

2.3.3.2. *Caves and sinkholes*

In the karst regions of New Zealand, numerous underground cavities yielded moa remains (Figure 2.3). Most of the time, these cavities worked as traps that the birds, once they had entered them or fallen into them, could not leave. Such karst sites have been known since the 1850s, when Hochstetter explored a cave with moa bones in the valley of the Aorere River. Since then, many excavations have been carried out in caves and sinkholes, in particular since the 1990s (Worthy & Holdaway, 2002).

2.3.3.3. *Alluvial deposits*

Sediments laid down by rivers sometimes contain moa bones. The first bones of these birds that were discovered, as early as the 1830s, were from such sedimentary environments (Oliver, 1949). Footprints and eggs were also discovered in fluvial deposits (Figure 2.3).

2.3.3.4. *Swamps*

The largest accumulations of moa bones came from swamps (Figure 2.3). They were generally found in blackish peat levels rich in plant remains, overlying bluish clay. Sometimes, the distal bones of moa limbs were preserved in a vertical position, which suggests that the birds were mired in the swamp mud. The very productive sites of Glenmark, Hamilton, Te Aute, Kapua, Enfield and Pyramid Valley, to mention but a few, correspond to swamps often established at the location of an old lake. Sometimes, thousands of bones were collected from such deposits, corresponding to hundreds of individuals (approximately 800 from Kapua, 800 to 900 from Enfield: Oliver, 1949; Anderson, 1989).

2.3.3.5. *Hunting sites*

We can gather under this umbrella term sites where moa remains were accumulated by human activities, in this case by the Maoris, who hunted and consumed these birds. Bones (and eggs) were then part of kitchen waste (midden) accumulations along with stone tools and other implements, as well as the remains of other hunted (seals) or domestic (dogs) animals. Many such sites are known from New Zealand and are often located along the coast and along rivers (Anderson, 1989; Berentson, 2012) (Figure 2.3). They reflect large-scale moa exploitation by the Maoris, which quickly led to the disappearance of these birds. Hunting camps, like those of Shag Point, Awamoa and Waitaki River, have been reconstructed, where moa, sometimes after being transported by water on rafts, were skinned and then cooked in stone ovens especially designed for this purpose, before being consumed (Berentson, 2012).

2.4. Anatomy

Figure 2.4. *Skeleton and reconstruction of a male* Dinornis robustus

2.4.1. *Skeleton*

Dinornithiformes form a relatively varied group in terms of size and body proportions (Figure 2.4), but still present a certain morphological unity (Figure 2.5) (for detailed anatomical descriptions, see Worthy & Holdaway (2002) and Worthy & Scofield (2012)).

The head was small compared with the body (Figure 2.5) and the beak was short, more or less pointed depending on the species and slightly curved. The mandible had a clearly concave ventral profile in lateral view. The mandibular symphysis was short.

The neck was long and composed of rather short cervical vertebrae. Former reconstructions of moa showed them with an extremely erect neck and a rather vertical cervical column. Later studies of the intervertebral articulations showed that the moa posture was different, as the spine was in a more horizontal position, and the neck formed a more accentuated ventral curve (Worthy & Holdaway, 2002) (Figure 2.4).

A noticeable trait of moa was the complete disappearance of wing bones. The pectoral girdle remained present, under the form of a single bone formed by the fusion of the coracoid and scapula (scapulocoracoid), but the bone elements of the upper limb had totally disappeared.

The pelvis was robust and wide, which differentiated it from that of the cursorial ratites, which have a very elongated and very narrow pelvis.

The hindlimbs were powerful (Figure 2.5). The femur was short, the tibiotarsus very long, and the tarsometatarsus more or less short and robust according to the species (see below, section "locomotion"). They had four toes. The ungual phalanges were pointed and slightly curved.

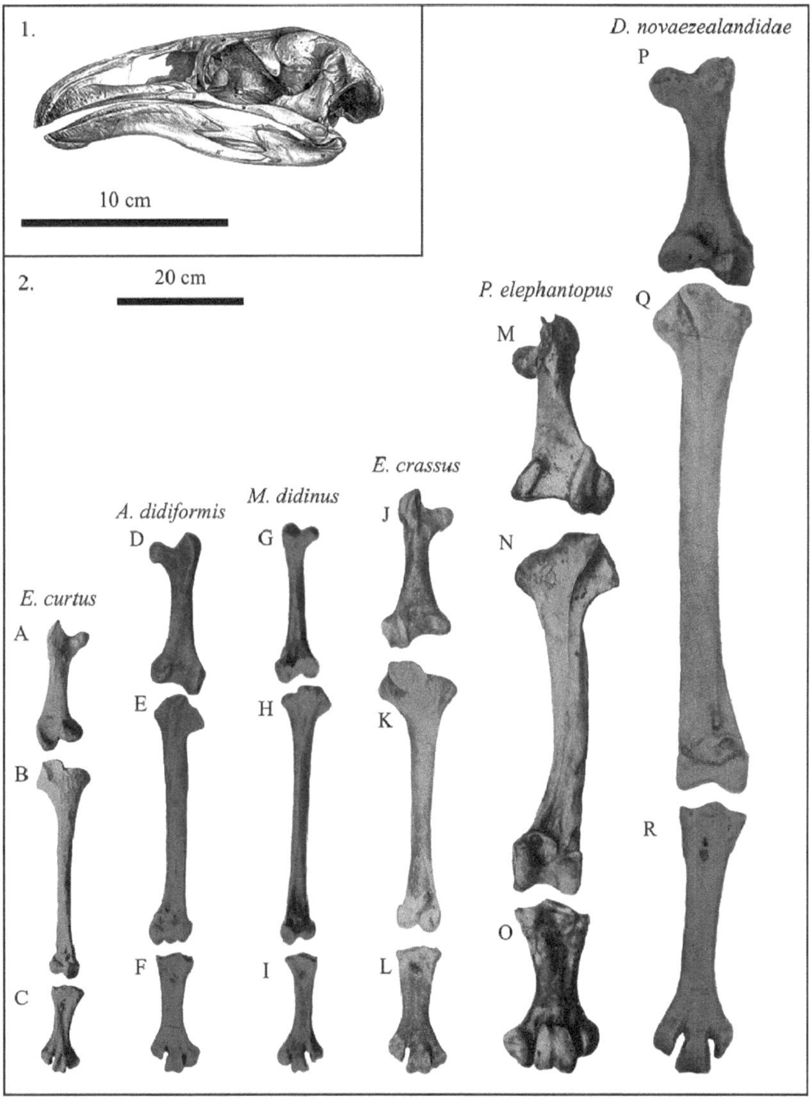

Figure 2.5. *(1) Skull of* Dinornis novaezealandiae *from Owen (1879a); (2) Diversity of the hindlimbs bones of Dinornithiformes: A, D, G, J, M, P: femur; B, E, H, K, N, Q: tibiotarsus; C, F, I, L, O, R: tarsometatarsus, from Oliver (1949)*

2.4.2. *Soft parts*

As these birds have recently become extinct, and thanks to some sites with exceptional preservation conditions, we know of numerous cases of soft part preservation in moa. Thus, mummified specimens were discovered in caves of the South Island, with muscles, skin and feathers all preserved (Figure 2.6). A notable example of this is the type specimen of *Megalapteryx didinus*, found in a cave in the Queenstown district in 1876, which still had a great part of its skin and feathers, and showed that this moa had feathers from the base of the beak to the base of the toes (Oliver, 1949). Cartilaginous elements, like tracheal rings, can also be preserved.

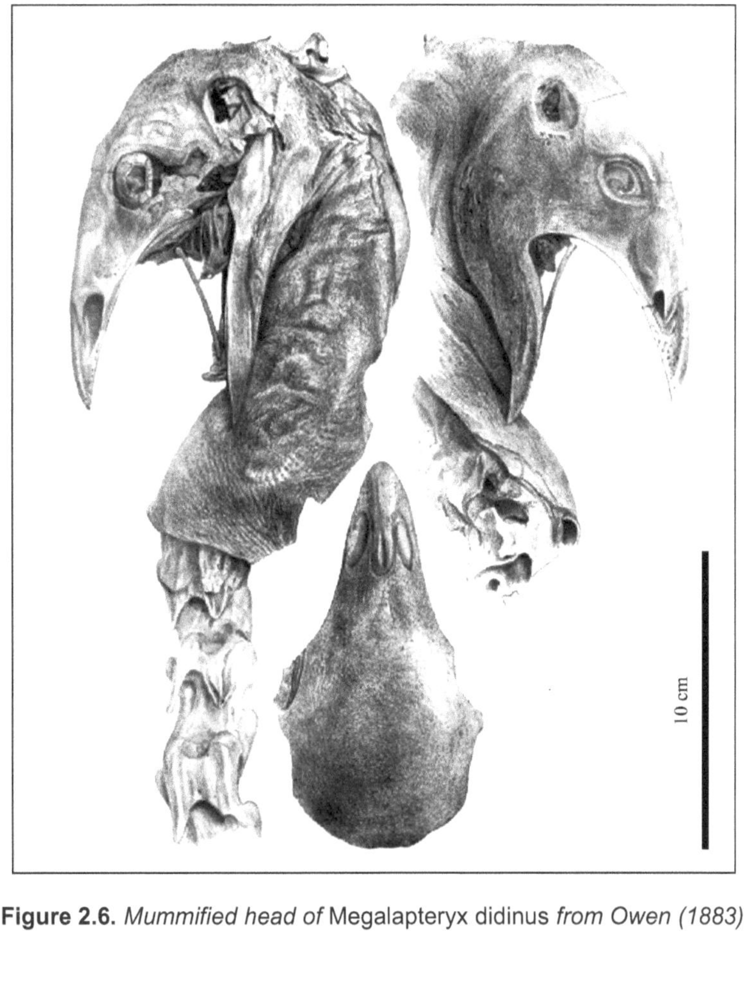

Figure 2.6. *Mummified head of* Megalapteryx didinus *from Owen (1883)*

Many isolated moa feathers were collected from the sediments of rock shelters and in the silt of rivers, and enabled reconstructions of the plumage of these birds. The feathers were relatively short, filamentous, flexible and without barbicels, sometimes with a double shaft. The original color of the feathers was often preserved in rock shelters under a dry climate, and the study of mitochondrial DNA made it possible to ascribe feather types to definite moa species (Rawlence *et al.*, 2009). Two main types can be distinguished: light brown at the base becoming darker up to black at the tip (e.g. in *Dinornis robustus*), and dark brown at the base becoming white at the tip (in *Pachyornis elephantopus*).

2.5. Paleophysiology and paleoecology

2.5.1. *Size and body mass*

The very large size of some moa surprised the first researchers who were interested in these birds, but there were also smaller species. It should be noted that moa height estimations were particularly overestimated in the past, at a time when they were reconstructed in an erect position, with a very vertical spine. Current reconstructions, with a more horizontal spine, inevitably result in lower height estimations (Table 2.1). Bunce *et al.* (2009) provided the following information for the six currently accepted moa genera.

	Height (cm)	Body mass (kg)
Dinornis	90–200	56–249
Megalapteryx	65–95	28–80
Emeus	73–99	36–79
Euryapteryx	51–103	12–109
Anomalopteryx	50–90	26–64
Pachyornis	54–121	17–163

Table 2.1. *Height and body mass estimations of Dinornithiformes at the genus level, according to Bunce* et al. *(2009)*

The large variation magnitudes are explained by the fact that all these genera, apart from *Megalapteryx*, include several species with different sizes and by the fact that there is a very marked sexual dimorphism in some of them.

At the species level, Worthy and Scofield (2012), using Dickison's results (2007), provided the following mass estimations (Table 2.2), which were calculated from the femur circumference, while emphasizing that more work was required to achieve reliable results.

	CI + 95%	CI – 95%	Average (kg)
Dinornis robustus	345.5	164.0	238.0
Anomalopteryx didiformis	59.3	40.4	49.0
Megalapteryx didinus	25.2	18.2	21.4
Emeus crassus	144.2	83.4	109.6
Euryapteryx curtus	47.8	33.8	40.2
Euryapteryx gravis	156.7	86.1	116.2
Pachyornis australis	156.7	86.1	116.1
Pachyornis elephantopus	322.1	147.4	217.9
Pachyornis geranoides	43.2	30.9	36.5

Table 2.2. *Body mass estimations of some species of Dinornithiformes, according to Worthy and Scofield (2012) and Dickison (2007) (CI = Confidence Interval)*

Therefore, it should be noted that there are differences between the results offered in the various publications due to the use of different methods.

2.5.2. *Sexual dimorphism*

A very significant reversed sexual dimorphism was discovered in some moa thanks to the study of their DNA, which made it possible to differentiate sexes (Bunce *et al.*, 2003; Huynen *et al.*, 2003) and confirm previous assumptions. It was thus shown that the female was larger than the male in *Dinornis*, *Euryapteryx curtus*, *Emeus crassus* and *Pachyornis geranoides*. This dimorphism was especially striking in *Dinornis*, in which the mass of the largest females was up to 280% of that of the largest males (150% for height) (Bunce *et al.*, 2003). Such a significant dimorphism is unique among birds and land mammals.

The excessive multiplication of Dinornithiformes taxa by former authors is partly explained by this dimorphism, the existence of which was only demonstrated recently.

2.5.3. *Locomotion*

Dinornithiformes, which were completely wingless, were purely land birds. As mentioned above, the recent reconstructions of moa showed that they had a rather horizontal back and a sigmoid curve of the neck. This posture would be suitable for locomotion in environments with dense vegetation (Worthy & Holdaway, 2002).

Several species of moa were characterized by very sturdy limbs with short and wide tarsometatarsi. This was especially true for Emeidae, and even *Dinornis*, which had more slender limbs, was part of the graviportal forms according to the locomotion-type estimation method proposed by Storer (1960). Using the method proposed by Angst *et al.* (2016; Figure 2.7), based on tarsometatarsus proportions, it appears that all the moa studied (*Dinornis robustus*, *Dinornis novaezealandiae*, *Anomalopteryx didiformis*, *Pachyornis elephantopus*, *Pachyornis australis*, *Pachyornis geranoides*, *Emeus crassus*, *Euryapteryx gravis* and *Euryapteryx curtus*) are of the graviportal type, with length/width ratios smaller than 12, even though this is less pronounced in *Dinornis* than in Emeidae.

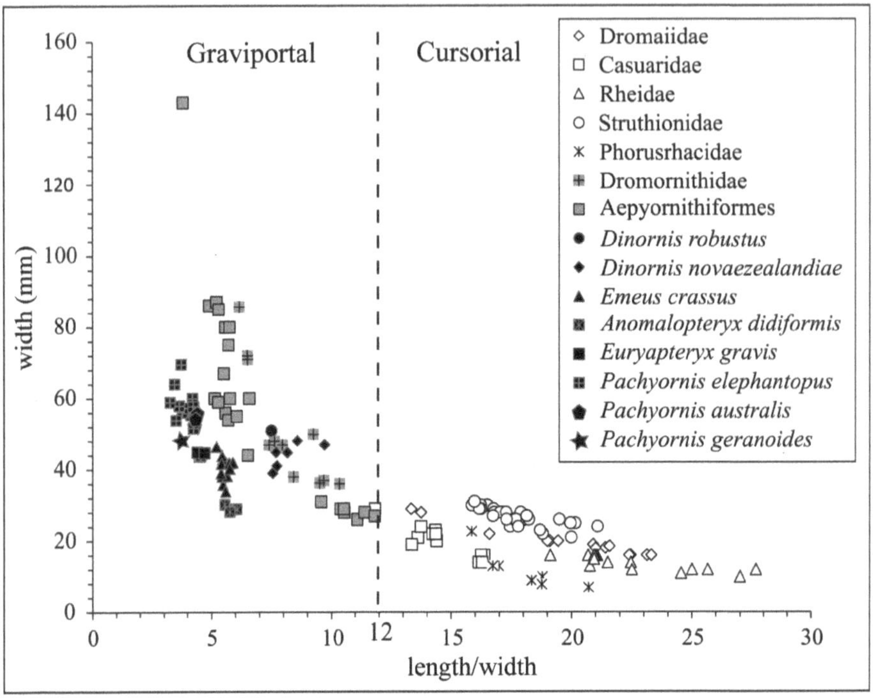

Figure 2.7. *Estimation of the locomotion type of Dinornithiformes from the method of Angst* et al. *(2016)*

The graviportal condition becomes extreme in *Pachyornis elephantopus*, which was the subject of a special study by Alexander (1983). According to this author, the robustness of the limb bones of this moa is not disproportionate compared with its mass.

To estimate the walking speed of moa, their footprints have been used (Figure 2.8). According to Worthy and Holdaway (2002), only 6 or 7 moa trackways are known. Some sites record the passage of several individuals. Alexander (1983) published speed estimations based on several of these tracks and they vary from 0.8 to 2 m/s. These are rather low speeds, indicating walking rather than running. Worthy and Holdaway (2002) obtained slightly different speeds from the same trackways, because their estimations of the birds' height at the hip were not exactly the same, but the results obtained, between 3.84 and 4.34 km/h, also indicated a slow movement. However, it is important to keep in mind that speeds estimated from tracks do not generally correspond to the maximum speed that could be reached by the animal that made them. The graviportal nature of all the moa suggests, however, that these birds were not fast runners.

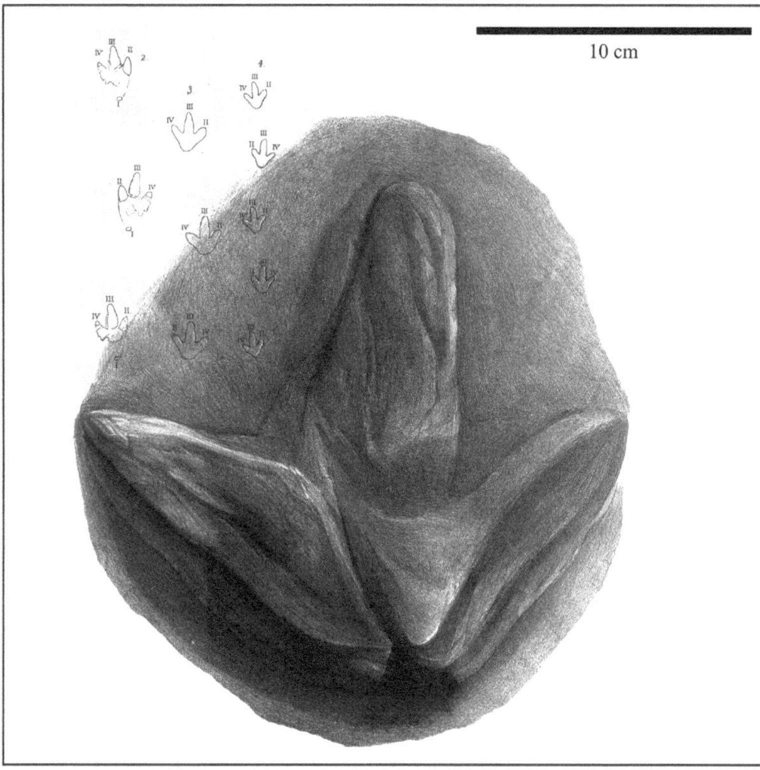

Figure 2.8. *Drawing of a moa footprint, from Owen (1879a)*

2.5.4. *Diet*

The shape of the jaws of the Dinornithiformes indicates that these birds were herbivorous. Gastroliths were found in abundance in connection with their skeletons, especially in swamps. Stomach contents and coprolites helped to identify their diet. The stomach contents of *Dinornis*, *Emeus* and *Euryapteryx*, in particular from the Pyramid Valley swamp dating from approximately 3000 years B.P. (Before Present) studied by Burrows (1980), indicated that these moa mainly ate twigs of woody plants, thus showing that these birds found their food in a forest environment.

Numerous coprolites are known from the late Quaternary of New Zealand (Wood & Wilmshurst, 2014) and it is now possible to more accurately identify them owing to the DNA they contain, just as plant DNA found in coprolites allows us to more fully identify the plants consumed (Wood *et al.*, 2013). The study of coprolites of Dinornithiformes thus provided very significant information on the diet of moa. It turns out that moa were more generalist than previously thought. Wood *et al.* (2008) thus demonstrated that four species of moa (*Dinornis robustus*, *Megalapteryx didinus*, *Pachyornis elephantopus* and *Euryapteryx gravis*) that lived on the South Island during the Holocene consumed various plants (more than 30 taxa), predominantly herbs and low shrubs, with no pronounced differences between the diets of these different sympatric species. Nevertheless, *Megalapteryx didinus* seems to have consumed more herbs, while *Dinornis robustus* also ate tree leaves. A more recent work by Wood *et al.* (2013) specified the diets of four species (*Dinornis robustus*, *Anomalopteryx didiformis*, *Pachyornis elephantopus* and *Megalapteryx didinus*) that lived during the four centuries before the extinction of the group, around 1450 A.D. The greatest differences were between *Pachyornis elephantopus*, which mainly ate herbs, and *Anomalopteryx didiformis*, which mainly consumed forest plants. *Dinornis robustus* and *Megalapteryx didinus* had more varied diets and habitats.

Zinoviev's work (2013) on moa limb myology, including *Emeus crassus* and *Dinornis robustus,* suggested *inter alia* that the second toe of their foot may have been particularly used to scrape and dig, in order to search for food such as rhizomes.

Attard *et al.* (2016) used finite element analysis to study the stress exerted on the beaks of five moa genera, and to draw conclusions from it regarding their diet. Among the results, it can be noted that *Euryapteryx curtus* had a beak that was weaker than those of the other moa and probably ate mainly fruits and soft leaves. The other Dinornithiformes studied were better at grasping and cutting fibrous leaves and twigs. *Pachyornis australis* allegedly preferred lateral head movements, whereas *Anomalopteryx didiformis* was better at performing back and forth movements. *Megalapteryx didinus* and *Dinornis novaezealandiae* were thought to be well suited to twisting movements. These different practices to acquire food can explain how various species of moa shared the same environments.

2.5.5. *Reproduction and growth*

2.5.5.1. *Eggs*

Gill (2006) published a catalog of moa eggs that were sufficiently well preserved to provide significant measurements, 36 in total, including 23 from "natural" sites (dating from 1300 to 19000 years B.P.) and 13 from archeological sites, sometimes graves (dating from 400 to 700 years B.P.; Huynen *et al.*, 2010). Fragments are evidently more abundant. According to Gill (2007) and Huynen *et al.* (2010), the smallest eggs were 10 times the volume of a hen's egg (120 x 95 mm), the largest ones 80 times that volume (240 x 178 mm) (Figure 2.9.). Egg color ranges from olive green to white and shell thickness from 0.5 to 1.89 mm (Gill, 2007; Huynen *et al.*, 2010).

Figure 2.9. *A Moa egg (left) compared with a hen's egg (right), from Owen (1879a)*

As for many other aspects of moa biology, the study of ancient DNA has led to great progress concerning their eggs, as they can now be ascribed to definite species. Huynen *et al.* (2010) thus succeeded in identifying eggs belonging to the six currently recognized moa genera. The presence of male DNA on the external surface of some eggs of *Pachyornis*, *Euryapteryx* and *Dinornis* suggests that the eggs were incubated by the males. Among extant ratites, eggs are incubated by the less heavy sex (most often the males), to reduce the risk of breaking them. It is not surprising that among moa, which displayed a significant reversed sexual dimorphism, males incubated the eggs. Huynen *et al.* (2010) pointed out that the eggshell of *Dinornis* was extremely thin (1 to 1.4 mm) for a bird that size, even though the incubation was performed by the male (weighing approximately 75 kg), and that consequently the risks of breaking the eggs were greater than those among most known birds. It is possible that eggs were laid in nests specifically built to bear the weight of the brooding bird, but moa nest structure remains very poorly known, even though Wood (2008) described the materials (twigs, bark fragments) used by moa to build their nests, preserved in rock shelters in a dry climate. These materials suggest that nests were similar to those of the extant ratites of Australasia (emu and cassowary).

2.5.5.2. *Embryonic development and growth*

As early as the 19th Century, a moa embryo preserved inside an egg discovered on a riverbank in Cromwell, on the South Island, was described by Hector (1872). The morphology of the premaxillary made it possible to ascribe it to *Pachyornis*. However, it is the study of bone histology that mostly contributed to our knowledge of moa growth. A study by Turvey *et al.* (2005) revealed the presence of lines of arrested growth in the cortical bone of Dinornithiformes, indicating a slower growth than for most birds, which reach their adult size and sexual maturity in approximately a year. *Dinornis* required at least 3 years to reach it, *Megalapteryx* up to 5 years, *Anomalopteryx* up to 8 years and *Euryapteryx* possibly up to 9 years. It suggests a K reproductive strategy for Dinornithiformes, with a long life expectancy, and an infrequent and late reproduction, which would reflect an adaptation to the very particular insular environment of New Zealand. With no mammalian predators, *Dinornis*, the largest moa, had a different type of growth than the smaller Emeidae, and reached its large size through a faster development and not by extending its growth duration. The study of *Dinornis* bones from swamps, considered as corresponding to real moa populations, indicates that 72.5 to 87.3% of the individual specimens had reached their adult body mass and that 55.9 to 78.2% had reached sexual maturity (Turvey & Holdaway, 2005). The percentage of adults is therefore high, which is consistent with a K reproductive strategy. Moa life expectancy was probably high, possibly around fifty years (Worthy & Holdaway, 2002).

Huynen *et al.* (2014) identified a series of immature moa bones by means of their DNA. This allowed them to study growth in *Euryapteryx* (one of the specimens apparently belonging to a recently hatched chick) by using morphometric, isotopic and radiographic techniques. One of their conclusions was that moa chicks must have been greatly dependent on adults for a long time because of their slow growth, unlike extant ratites, which are precocial and do not depend on their parents for a long time. This peculiarity may be linked to the absence of land predators in New Zealand before the arrival of humans.

2.5.6. *Paleopathology*

We know relatively little about the pathologies that affected moa (Worthy & Holdaway, 2002). Some cases of fracture are known. Severe arthritis cases were reported, especially relating to the phalangeal joints. Cases of chronic osteomyelitis resulting in serious abscesses of limb and mandible bones are also known.

2.6. Paleoenvironment

2.6.1. *Climate and vegetation*

It was long believed that moa mainly lived in open environments such as grasslands, rather than in forest environments. This idea was linked to a misconception of the environments of New Zealand at the time when these birds lived there. Grasslands covered with herbs or ferns over large areas in some regions when Europeans arrived did not correspond to natural vegetation. It was from the 1960s onwards that the idea according to which moa instead lived in forest environments gradually took over, in particular thanks to pollen analyses, which showed that grasslands were not very developed during the Holocene. Works on moa diet (see above) confirmed that they did not only frequent open habitats. However, abundant moa remains were also found in the subalpine zone, up to 1800 meters above sea level, above the tree line (Worthy & Holdaway, 2002). Therefore, it now appears that moa lived in rather varied environments. Owing to the large number of known sites and the abundant remains of these birds, it is possible to provide distribution maps indicating the different species and their respective abundances in different parts of New Zealand, according to the types of environment found there. As indicated by Worthy and Scofield (2012), some moa lived in extremely varied environments. Thus, both species of *Dinornis*, *D. novaezealandiae* on the North Island and *D. robustus* on the South Island, frequented all the vegetation zones, from coastal to subalpine areas. Other species, such as *Megalapteryx didinus*, which mainly lived in the highlands of the South Island, had less vast distributions. Within the same genus, different species could have different preferences in terms of habitat. For example, *Pachyornis elephantopus* mainly frequented the low regions of the South Island, whereas *Pachyornis australis* preferred subalpine areas. However, as shown by faunal lists, in numerous regions, several moa species coexisted in the same environment. The above-mentioned studies on the different diet adaptations in various species help us to better understand these cohabitations, by adaptation to distinct ecological niches in the same territory.

The environments in which moa lived changed over time, especially during the Quaternary glaciations that affected New Zealand and led to substantial changes in vegetation. As mentioned above, many species known during the Holocene were already present at older periods of the Pleistocene, and the climate changes of the Quaternary apparently did not result in significant extinctions among Dinornithiformes, even though glaciations clearly influenced speciation phenomena within the group (Bunce *et al.*, 2009). However, size modification phenomena over time can be noted. According to Worthy and Holdaway (2002), several moa species saw their size reduced between the glacial maximum of the Late Pleistocene and the Holocene. Thus, the average body mass of *Pachyornis geranoides* reduced from 41 to 27 kg, that of *Euryapteryx curtus* from 24.5 to 20 kg and that of *Megalapteryx didinus* from 56 to 40 kg. The case of *Pachyornis elephantopus* is particularly interesting, because an increase in size as the climate cooled is apparent, with a maximum during the coldest period, followed by a decrease when the climate warmed up. The population that lived during the last glacial maximum was 34% heavier than the population of the late Holocene. This may be an application of Bergmann's rule, according to which within a zoological group, forms living in a cold climate are larger than those living in a warmer climate.

Interactions between moa and the plant environments they frequented were the subject of in-depth studies, and it was suggested that co-evolution phenomena occurred. According to Atkinson and Greenwood (1989), at least 11 types of growth characteristics are believed to have evolved among indigenous plants in response to consumption by moa, resulting in less attractive plants. These include in particular the development of thorns in the genus *Aciphylla*, camouflage phenomena and the evolution of divarication among some plants, whose branches are divided at very open angles to form bushes with small leaves that are hard to access (Greenwood & Atkinson, 1977).

2.6.2. *Predators*

Before the arrival of humans in New Zealand, at a geologically recent date, moa had no land predators, in the absence of mammals or large carnivorous reptiles (the evolution and diversification of these large birds that were unable to fly was certainly made possible by this absence). It seems that the only predator that attacked Dinornithiformes was Haast's eagle, *Harpagornis moorei*, a large bird of prey now extinct that is believed to have weighed up to 10 kg for male specimens and 15 kg for females, with a wingspan that could reach three meters. Remains of this eagle were only found on the South Island. The fact that *Harpagornis moorei* attacked moa is proven by predation traces (Worthy & Holdaway, 2002): a number of moa pelves bearing traces of the eagle's claws were discovered, showing that this bird of prey attacked even the largest of these birds (*Dinornis*). The dorsal bony plate formed by the ilia was perforated by the claws, which penetrated the air sacs located in the kidney area, causing them to bleed to death. The prey of Haast's eagle was far heavier than it was, as it could weigh up to 200 kg. It is the highest ratio between the mass of the prey and that of the predator that is known among terrestrial vertebrates. Once a moa was killed, Haast's eagle could dismember it thanks to its powerful beak, its shape recalling that of a vulture rather than that of an eagle. The disappearance of *Harpagornis moorei* seems to have been linked to the extinction of the moa, and so to the arrival of humans in New Zealand.

2.7. Extinction

The date of the extinction of the moa and its causes have been the subject of numerous discussions and speculations since the 19th Century, when it became clear that these birds no longer existed (note however that the cryptozoological literature includes numerous reports about living moa sightings, up to the present day).

In the past, "natural" causes were proposed to explain moa extinction. The idea of an evolution leading towards gigantism, resulting in an inability to adapt to environmental changes, mentioned by Oliver (1949), belongs to views on "racial senescence", often discussed at the beginning of the 20th Century about dinosaurs, and only has an anecdotal value.

Climate change occurring since the last glacial period was also mentioned, especially by Archey (1941), as a factor that contributed to the decline of moa: open environments, supposedly frequented by these birds, were thought to have regressed in favor of forest environments, which were less favorable to Dinornithiformes. As noted earlier, modern views on the environments in which moa lived have significantly changed, and it is admitted today that they had varied habitats and food types. The hypothesis according to which climate change could have significantly contributed to the extinction of these birds is therefore not convincing.

The role of humans in the disappearance of moa has been discussed for a long time and is nowadays beyond doubt. On the other hand, the manner and the chronology of this extermination were the subject of long discussions. Numerous archeological sites show that moa had been hunted by humans on a large scale (see Anderson, 1989; Berentson, 2012). The first discoveries of such sites date back to the 1860s. At that time, Julius von Haast developed the theory according to which moa had been hunted, and then exterminated, by archaic populations, probably Melanesian, well before the arrival of the Maoris. The existence of this "moa hunting" Paleolithic type of culture was the subject of much controversy (see Anderson (1989) and Berentson (2012) for a history of this debate). Nowadays, the hypothesis of these archaic moa hunters prior to the Maoris has been abandoned, and it is clearly the latter, the first humans to have reached New Zealand, who were responsible for the extermination of the Dinornithiformes. Excavations carried out at many archeological sites revealed in detail how the giant birds were hunted, killed, butchered and cooked in ovens built for this purpose, in vast camps often located at river mouths (which helped with transporting moa carcasses on rafts). To direct predation was added the destruction of habitats, especially forests, by fire.

According to the most recent and most reliable archeological data and carbon-14 dating, the arrival of the Polynesians in New Zealand took place during the 13th Century A.D. (Wilmshurst *et al.*, 2011), probably between 1230 and 1260 A.D. (Perry *et al.*, 2014). Moa were then quickly exterminated: according to Perry *et al.* (2014), hunting is believed to have stopped (due to lack of game) approximately 150 years after the beginning of the colonization by the Maoris, and complete extinction is believed to have taken place approximately fifty years later. Moa would then have disappeared around the middle of the 15th Century. The different moa species are believed to have disappeared approximately at the same time, regardless of their size (Perry *et al.*, 2014). The disappearance of moa, mainly due to overkill, even though the destruction of habitats might have contributed to it, would then have been abrupt and quick. It is however still difficult to accurately date the disappearance of the last moa. Some authors assume that some species survived in remote mountainous regions up to a relatively recent date, maybe even up to the colonization by Europeans (see in this respect Berentson (2012)). Stories about moa survival up to the present day are however not confirmed, even though it is one of the favorite topics of cryptozoologists.

3

Aepyornithiformes

3.1. Historical background

Apart from the tales about the Roc bird (Buffetaut, 2013a), which may actually be unrelated to Aepyornithiformes (Decary, 1937), the first account to mention these giant Malagasy birds seems to belong to Flacourt (1661) who, in his *Histoire de la grande isle de Madagascar*, calls *Vouron patra* a large bird "which lays eggs like an ostrich" and is "a species of ostrich" that lives in the most deserted places of the southern part of the island. This short note suggests that Aepyornithiformes were still alive in the 17th Century.

The first concrete proof of the past existence of giant birds in Madagascar was pieces of eggshells brought back to France by travelers in the 1830s (Lavauden, 1931). The first estimates of the volume of the eggs, based on fragments, were too low, and it was only in 1851 that Isidore Geoffroy Saint-Hilaire could present to the *Académie des Sciences* in Paris huge entire eggs obtained by a certain Captain Abadie and whose volume corresponded to that of 148 hen's eggs (Geoffroy Saint-Hilaire, 1851). As some bones, among which a tarsometatarsus, indicated that this was a bird, Geoffroy Saint-Hilaire named it *Aepyornis maximus* and estimated its height at around 3.80 meters. Even though he considered *Aepyornis* similar to ostriches, other naturalists disagreed. In particular, Bianconi, an Italian paleontologist influenced by Marco Polo's writings about the Roc bird, decided that it was an enormous bird of prey (Bianconi, 1865). After the excavations carried out at Ambolisatre (or Ambolisatra), in the south-west of the island, by an explorer called Alfred Grandidier, Milne-Edwards and Grandidier (1869) were able to show that *Aepyornis* was actually a ratite on the basis of a certain number of limb bones.

In the last decades of the 19th Century, while Europeans kept exploring Madagascar, important fossil sites were discovered (especially in Antsirabe, in the center of the island) and a large number of specimens were sent to Europe, where various museums nowadays keep vast collections of the remains of Aepyornithiformes. The genus *Mullerornis*, smaller and slenderer than *Aepyornis*, was recognized by Milne-Edwards and Grandidier (1894). In this period, a large number of species were described. Monnier (1913) published a monograph on *Aepyornis* where he attempted to reduce this number. A similar work was carried out by Lamberton (1934) on *Mullerornis*.

At the beginning of the 1930s, with the diversification of the specimens of Aepyornithiformes discovered and a larger number of descriptions and studies, Lambrecht (1929) suggested that the species *Stromeria fajumensis* (now *Eremopezus eocaenus*) was the African continental ancestor of Aepyornithiformes. However, this hypothesis raises a problem. *Stromeria fajumensis* was a large land bird known to have lived at the end of the Eocene in Egypt (Rasmussen *et al.*, 2001). If this land bird is actually the ancestor of Aepyornithiformes, we must then admit that it swam across the Mozambique Channel at the beginning of the Tertiary. Even if this kind of dispersal has been observed for other groups, it is now being widely debated (Rasmussen *et al.*, 2001; Mayr, 2009). The second hypothesis is that there is no kinship between *Stromeria fajumensis* and Aepyornithiformes and that the ancestors of Aepyornithiformes were flying birds that lived at the beginning of the Tertiary and flew to the island (Rasmussen *et al.*, 2001; Mayr, 2009). Currently, the issue of the origin of Aepyornithiformes is still being widely debated and has not yet been solved.

Similarly, the systematics, paleoecology and paleobiology of these birds are still poorly understood. Despite some significant works, such as Wiman and Edinger's study on the brain of Aepyornithiformes (1940) and Balanoff and Rowe's description of an *Aepyornis* embryo inside an egg (2007), relatively little research has been carried out on these birds in the last few decades, and in consequence they are significantly less well known than other extinct giant birds such as moa in New Zealand.

3.2. Taxonomy, systematics and classification

3.2.1. *Taxonomy*

The taxonomy of Aepyornithiformes is particularly chaotic. The first taxonomic articles were poorly illustrated and descriptions were most often very limited. Assigning specimens to this group is in most cases linked with geographical considerations, considering that all large birds found in Madagascar are necessarily linked to this order (Balanoff & Rowe, 2007). The taxonomy of Aepyornithiformes has changed several times over the past 150 years. Geoffroy Saint-Hilaire was the first to mention this group of giant birds (1851), on the basis of remains found in Pleistocene and Holocene sediments in Madagascar, which he called *Aepyornis maximus*. However, on this occasion he did not erect the family Aepyornithidae, which was erected later by Bonaparte (1853), just as the order Aepyornithes (Aepyornithiformes according to current nomenclature) was erected only 30 years later by Newton (1884). After the description of *Aepyornis*, the additional genus *Mullerornis* was erected and included in Aepyornithidae by Milne-Edwards and Grandidier (1894) on the basis of its more slender legs. Three species were described in this genus based on very limited anatomical features: *Mullerornis betsilei*, *Mullerornis rudis* and *Mullerornis agilis* (Milne-Edwards & Grandidier, 1894). Finally, after reviewing all the previous works on *Mullerornis*, Lamberton (1934) confirmed the existence of the three above-mentioned species and suggested that a fourth one, *Mullerornis grandis*, should be added, by re-determining clearly and precisely the anatomical features that made it possible to define each of these species. These four species are currently considered valid.

In his taxonomic review, Lamberton (1934) examined all the material referred to the genera *Aepyornis* and *Mullerornis*, all of them grouped in a single family, Aepyornithidae. His meticulous study of the specimens allowed him to propose the erection of two families in the order Aepyornithiformes, Aepyornithidae and Mullerornithidae, each including a single genus, *Aepyornis* and *Mullerornis*, respectively. He put forward this new classification by relying on a large number of anatomical features that he named and described. Surprisingly, all the works that followed his study did not take into consideration this new classification and kept grouping the genera *Aepyornis* and *Mullerornis* in the family Aepyornithidae. However, the classification put forward by Lamberton seems correct and should thus be used, pending a taxonomic revision of this group.

Before Monnier's work (1913), the taxonomy of the genus *Aepyornis* was especially chaotic, including the description of around 11 species on the basis of a relatively small number of fragmentary specimens. In 1869, based on such fragmentary specimens, Milne-Edwards and Grandidier (1869) described three species of *Aepyornis*, *A. maximus*, *A. medus* and *A. modestus*, then Capellini (1889) concluded that there were two species, *A. maximus* and *A. grandidieri*, based only on the study of eggs. In 1893, Burckhardt (1893) published a study that described a new species, *A. hildebrandti*, which was smaller in size than the other species described and identified on the basis of a very large number of anatomical features. In 1894, Andrews (1894) erected the species *A. titan* by studying only two bones (a femur and a tibiotarsus), whereas in the same year Milne-Edwards and Grandidier (1894) erected four new species, *A. ingens*, *A. cursor*, *A. lentus* and *A. mulleri*, based on very vague criteria. Most of these species were identified on the basis of their size (occasionally even mixing adult with juvenile specimens) and very limited anatomical features. This is why Monnier (1913) reconsidered all of the literature available in his time and suggested that these 11 species should be grouped into three species, after carrying out a study on the proportions and size of the bones (especially of hindlimb bones). These species are *Aepyornis maximus* Geoffroy Saint-Hilaire, 1851, *Aepyornis medius* Milne-Edwards & Grandidier, 1869, and *Aepyornis hildebrandti* Burckhardt, 1893, to which he added a new species, *Aepyornis gracilis* Monnier, 1913 (based on only one femur which was clearly less heavy and more slender than that of the other species). Therefore, the species *A. titan* Andrews, 1894, and *A. ingens* Milne-Edwards & Grandidier, 1894, were synonymized with *A. maximus* Geoffroy Saint-Hilaire, 1851 (Monnier, 1913). The species *A. medius* Milne-Edwards & Grandidier, 1869, includes the species *A. lentus* Milne-Edwards & Grandidier, 1894, and *A. cursor* Milne-Edwards & Grandidier, 1894, as well as a femur attributed to *A. maximus* by Geoffroy Saint-Hilaire, 1851 (Monnier, 1913). The species *A. modestus* described by Milne-Edwards and Grandidier (1894) corresponds to a specimen of the genus *Mullerornis* (Monnier, 1913). Finally, the species *A. grandidieri*, described on the basis of only one egg, is considered a *nomen dubium*, and the species *A. mullleri* Milne-Edwards & Grandidier, 1894, and *A. hildebrandti* are included in the species *A. hildebrandti* Burckhardt, 1893.

Nowadays, these four species are considered valid even though they are distinguished only on the basis of size differences. A global taxonomic review of Aepyornithiformes is therefore necessary in order to provide sound diagnoses for all these species.

Since most specimens in collections are not identified at the species level, most paleobiological and paleoecological studies have been carried out at the genus level. The genus *Mullerornis* is significantly slenderer than *Aepyornis*, so that they can be distinguished relatively easily. This is why in the rest of the chapter, all paleobiological and paleoecological considerations will be limited to genus names.

3.2.2. *Systematics*

The systematics of Aepyornithiformes is still quite chaotic. What is commonly acknowledged is that this order includes two families containing one genus each. However, the systematics of Aepyornithidae at the species level should still be revised as it is currently based only on differences in size (Monnier, 1913), so that the largest *Aepyornis* species corresponds to *A. maximus*, followed in a decreasing order by *A. medius*, *A. hildebrandti* and *A. gracilis*. Nowadays, the only systematics based on a precise study of aepyornithid specimens is that put forward by Monnier (1913) and shown here. The systematics of Mullerornithidae is slightly more precise thanks to Lamberton's review (1934), which described and compared all the specimens belonging to this group. Thus, Lamberton confirmed the three *Mullerornis* species erected by Milne-Edwards and Grandidier (1894) and added a fourth one, *M. grandis*.

Nonetheless, it is impossible to provide the specific stratigraphic and geographical occurrences of the different species for several reasons. First, as their taxonomy remains very vague and has changed several times, it is difficult to be sure whether the specific attributions of the specimens are correct. Moreover, a large part of the sites cannot be dated. Even when an excavation site has been recently dated, most specimens present in collections are not linked to precise information about their exact origin. Thus, it is in most cases impossible to assign a precise geographical and stratigraphic origin to a given specimen. This is reinforced by the fact that most of these specimens were collected during the 19th Century without any precise description or illustrations, so that fossils cannot be assigned *a posteriori* to a publication where geographical or stratigraphic information may have been provided. For these reasons, geographical and stratigraphic occurrences will only be provided at the level of the order Aepyornithiformes.

Class Aves Linnaeus, 1758

Super-order Palaeognathae Pycraft, 1900

Order Aepyornithiformes Newton, 1884

Stratigraphic occurrence: Pleistocene and Holocene of Madagascar.

Family Aepyornithidae Bonaparte, 1853

Genus *Aepyornis* Geoffroy Saint-Hilaire, 1851

– *Aepyornis maximus* Geoffroy Saint-Hilaire, 1851

– *Aepyornis medius* Milne-Edwards & Grandidier, 1869

– *Aepyornis hildebrandti* Burckhardt, 1893

– *Aepyornis gracilis* Monnier, 1913

Family Mullerornithidae Lamberton, 1934

Genus *Mullerornis* Milne-Edwards & Grandidier, 1894

– *Mullerornis betsilei* Milne-Edwards & Grandidier, 1894

– *Mullerornis rudis* Milne-Edwards & Grandidier, 1894

– *Mullerornis agilis* Milne-Edwards & Grandidier, 1894

– *Mullerornis grandis* Lamberton, 1934

3.2.3. *Classification*

The first to associate Aepyornithiformes with other groups was Valenciennes (1854), who hypothesized that they may be related to Auks and Penguins. Later on, Bianconi, who was convinced that *Aepyornis* corresponded to the Roc bird described in Arabian legends, tried to demonstrate that they were birds of prey and linked them to Falconiformes (Bianconi, 1870). It is now acknowledged that this family belongs to Paleognathae and more specifically to ratites (Monnier, 1913; Wiman, 1935;

Cracraft, 1974; Bledsoe, 1988; Pycraft, 1900; Lee *et al.*, 1997; Cooper *et al.*, 2001). However, there are very few studies focused on the position of Aepyornithiformes within ratites and the issue is still being debated. Cracraft (1974) describes this order as similar in many respects to kiwis and tinamous. More recently, Balanoff and Rowe (2007) considered it as the sister group of ostriches and rheas based on postcranial anatomical features, whereas the study carried out by Mitchell *et al.* (2014) on mitochondrial DNA regards it as the sister group of kiwis (Figure 3.1).

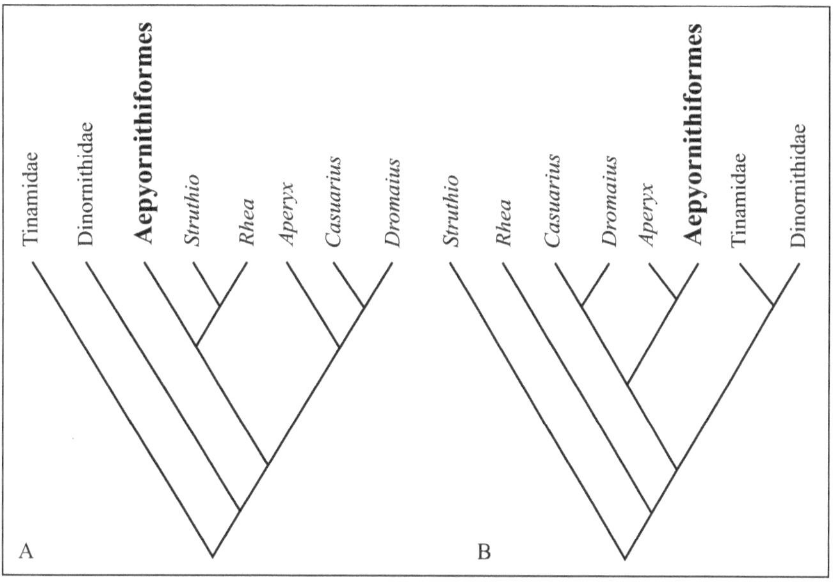

Figure 3.1. *The phylogenetic position of Aepyornithiformes according to the hypothesis put forward by Balanoff & Rowe (2007, A) and Mitchell* et al. *(2014, B)*

On the other hand, there is no study about the phylogenetic relations between the different families, genera and species of *Mullerornis* and *Aepyornis*. As of now, it is acknowledged that this order is monophyletic as all specimens come from the island of Madagascar, but this claim has never really been tested.

3.3. Stratigraphic and geographical distribution

Aepyornithiformes are known to have lived only on the island of Madagascar in the Pleistocene and Holocene (Monnier, 1913; Wiman, 1935) (Figure 3.2).

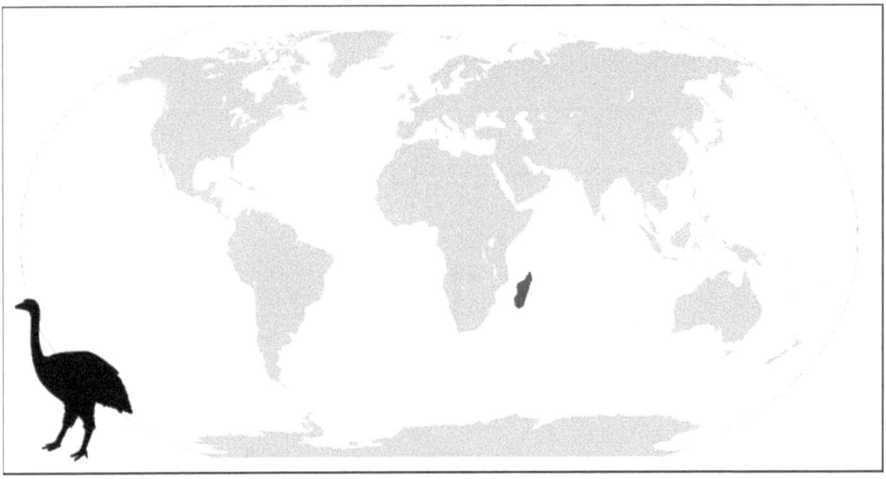

Figure 3.2. *Global geographic distribution of Aepyornithiformes. For a color version of the figure, see www.iste.co.uk/angst/birds.zip*

The oldest fossil remains attributed to these birds are eggshells found at Nasua and estimated to date back to around 22000 B.C., namely the end of the Pleistocene (Clarke *et al.*, 2006). The most recent recognized remains correspond to a fragment of eggshell of *Aepyornis* found at Fort Dauphin and estimated to date back to 1290 A.D. (Berger *et al.*, 1975; Burney *et al.*, 2004). These birds are thus known to have lived in Madagascar for at least 23000 years. They can be found in five specific areas of the island – north, center, west, south-west and south (Figure 3.3) and they are represented in the fossil record by both eggshells and bones. The geological and sedimentary environments where these remains have been found are mainly of two kinds: either volcanic flows that have blocked rivers and thus created marshes upstream where fossils have been preserved – this is especially the case for the highlands in the center of the island – or sand-dune environments that have built natural barriers and enabled the formation of marshes behind these ridges where specimens have been found (Mahé & Sourdat, 1972).

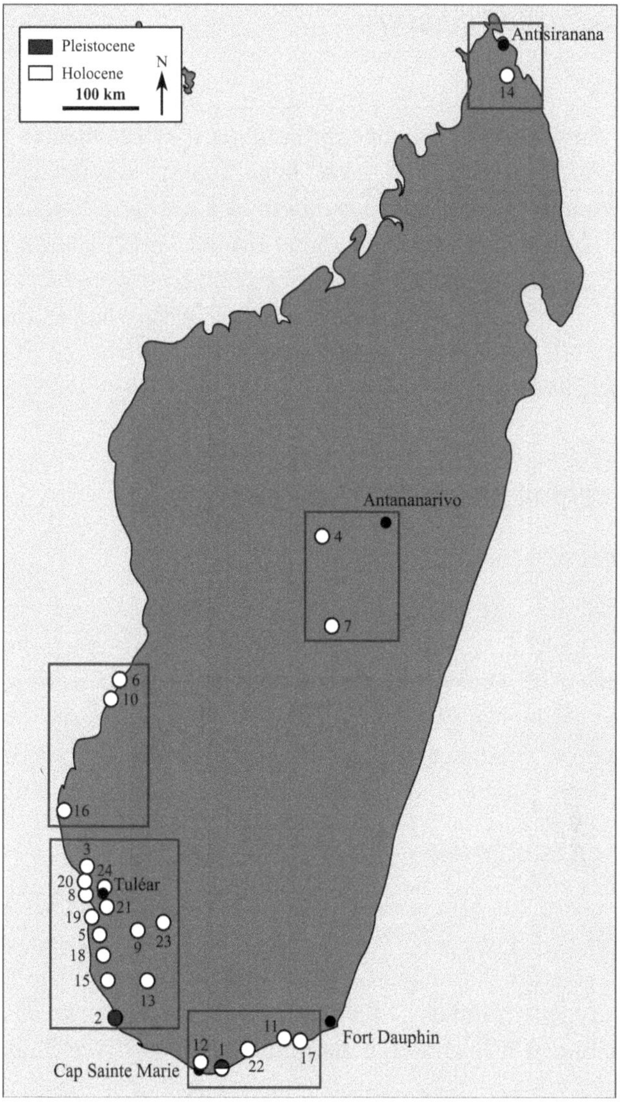

Figure 3.3. *Geographical distribution of the main fossil sites for Aepyornithiformes. Pleistocene sites: 1 – Faux Cap, 2 – Nasua. Holocene sites: 3 – Ambolisatra (or Andolonomby), 4 – Ampasambazimba, 5 – Anakao, 6 – Ankevo, 7 – Antsirabé, 8 – Belalanda, 9 – Belioky, 10 – Belo-sur-Mer, 11 – Beraroha, 12 – Cap Sainte-Marie, 13 – Ejeda, 14 – Irodo, 15 – Itampolo, 16 – Lamboharana, 17 – Maroaloke, 18 – Mitoho, 19 – Saint-Augustin, 20 – Songoritelo, 21 – Table Mountain, 22 – Talaky (or Manambovo), 23 – Taolambiby, 24 – Tuléar*

3.3.1. *Northern Madagascar*

3.3.1.1. *Irodo*

Irodo is the only site in the northern part of the island where fossil remains of Aepyornithiformes have been found. The excavation site is among the dunes close to Irodo Bay, south of the city of Antisiranana, where eggshells of *Aepyornis* have been found (Mahé, 1972; Clarke *et al.*, 2006) (Figure 3.3.). These eggshells have been estimated to date back to 680–1030 A.D. (Mahé & Sourdat, 1972; Burney *et al.*, 2004), whereas the excavation site itself overall dates back to 680–1260 A.D. (Mahé & Sourdat, 1972; Burney *et al.*, 2004).

3.3.2. *The central plateau of Madagascar*

3.3.2.1. *Antsirabé*

The Antsirabé site, also called Sirabé, lies east of the Ankaratra Mountains (Figure 3.3.) and represents one of the richest sites known to contain remains of Aepyornithiformes, having yielded a large number of bone remains (Milne-Edwards & Grandidier, 1894). This site corresponds to a former lake or wetland and dates back to between 22240 B.C. and now (Goodman, 1999; Burney *et al.*, 2004; Goodman & Jungers, 2013).

3.3.2.2. *Ampasambazimba*

This site lies in the Mazy valley, about 20 kilometers north of Itasy Lake (Monnier, 1913) (Figure 3.3.). It is the site where the only specimen referred to *Aepyornis gracilis* (a femur) has been found (Monnier, 1913), but also the origin of the type specimens of the species *Mullerornis grandis* (Lamberton, 1934) as well as of a fairly large assemblage of fossil vertebrates including many Aepyornithidae. Like Antsirabé, this is a former wetland or lake dating back to between 9850 B.C. and 1040 B.C. (Tattersall, 1973; MacPhee *et al.*, 1985; Burney *et al.*, 2004; Goodman & Jungers, 2013).

3.3.3. *Western Madagascar*

3.3.3.1. *Belo-sur-Mer*

The fossil site of Belo-sur-Mer also corresponds to the archeological site of Ankilibehandry (Burney *et al.*, 2004) (Figure 3.3.). This site dates back to between 370 B.C. and 960 A.D., and the remains of Aepyornithiformes date back to between 60 and 890 A.D. (Burney, 1999; Burney *et al.*, 2004). Belo-sur-Mer is also interesting as it is the site where the type of *Mullerornis rudis* was discovered (Lamberton, 1934; Goodman & Jungers, 2013).

3.3.3.2. *Ankevo and Lamboharana*

These two sites represent the other two sites known on the western coast of Madagascar (Figure 3.3). Only Lamboharana has been studied in detail and estimated to date back to between 790 B.C. and 990 A.D. (Mahé & Sourdat, 1972; Burney *et al.*, 2004).

3.3.4. *South-western Madagascar*

The south-western region of Madagascar is the one that contains the largest number of fossil sites that have yielded remains of Aepyornithiformes. This significant concentration has been interpreted as an area that presents numerous bodies of water, such as lakes or freshwater marshes, used as watering holes for all the fauna including Aepyornithiformes (Mahé & Sourdat, 1972).

3.3.4.1. *Tuléar and its region*

Tuléar (or Toliara) is the largest city in south-western Madagascar (Figure 3.3). Around Tuléar there are three fossil sites containing Aepyornithiformes: Balalanda, Songoritelo and Table Mountain. These three sites have provided fragments of eggshells that have been attributed to these large birds (Clarke *et al.*, 2006; Tovondrafale *et al.*, 2014). The Balalanda site, situated 6 kilometers from Tuléar, has been estimated on the basis of *Aepyornis* eggshells to date back to between 5210 ±140 years

B.P. which corresponds to around 4033 B.C. (Marden, 1967; Clarke *et al.*, 2006). The Songoritelo site lies around 15 kilometers from Tuléar and dates back to between 3261 and 5535 B.C. (Clarke *et al.*, 2006). Finally, Table Mountain, east of Tuléar, dates back to between 2235 and 5636 B.C. (Clarke *et al.*, 2006).

3.3.4.2. *North of Tuléar: Ambolisatra (or Andolonomby)*

The Ambolisatra (or Andolonobmy) site is situated 35 kilometers north of Tuléar and was discovered by Grandidier in 1868 and excavated from 1869 onwards (Gommery *et al.*, 2011) (Figure 3.3). It represents the first site containing subfossil vertebrates discovered in Madagascar (Mahé & Sourdat, 1972). This site has been estimated to date back to between 5900 B.C. and 410 A.D. (Burney *et al.*, 2004) thanks to several proxies. On the other hand, no dating is available for the Aepyornithiformes found at this site.

3.3.4.3. *The region south of Tuléar*

South of Tuléar, a large number of fossil sites containing Aepyornithiformes, among which are Saint-Augustin, Anakao, Itampolo, and Nasua (Figure 3.3), have been listed. Saint-Augustin deserves to be mentioned as it represents the original site of the complete eggs of *Aepyornis* described by Geoffroy Saint-Hilaire in 1851 and constituting the very first fossils attributed to these large birds (Geoffroy Saint-Hilaire, 1851; Monnier, 1913). The Anakoa site lies among coastal dunes estimated to date back to between 4840 and 2150 B.C. (Burney *et al.*, 2004) thanks to eggshells of *Aepyornis* eggs. The Itampolo site lies 2 kilometers north of the village of Itampolo and 200 kilometers south of Tuléar. The site itself dates back to between 150 B.C. and 1400 A.D., and an *Aepyornis* tibiotarsus with traces of human activity dates back to between 30 B.C. and 320 A.D. (Mahé & Sourdat, 1972; Burney, 1999; Burney *et al.*, 2004). Finally, the Nasua site (also called Nisua, Nisoa or Nishua) represents the oldest site where remains of Aepyornithiformes have been found. These remains include eggshells of *Aepyornis* that date back to the Pleistocene and more precisely to between 16742 and 21913 B.C. (Clarke *et al.*, 2006).

3.3.4.4. *The region south-east of Tuléar*

There are several fossil sites containing Aepyornithiformes south-east of Tuléar, among which are Ejeda, Belioky and Taolambiby (Figure 3.3). Of these, only the last one deserves to be described in some detail, as it represents one of the sites containing the oldest traces of human activity, which date back to between 416 and 365 B.C. (Burney *et al.*, 2004; Gommery *et al.*, 2011), whereas the site itself dates back to between 750 and 150 B.C. (Mahé & Sourdat, 1972; Burney *et al.*, 2004; Goodman & Jungers, 2013).

3.3.5. Southern Madagascar

3.3.5.1. *Talaky, Cap Sainte Marie, and Faux Cap*

Talaky, Cap Sainte Marie and Faux Cap are the three southernmost sites in Madagascar (Figure 3.3). The archeological site of Talaky also corresponds to the Manarnobo River Mouth paleontological site which dates back to between 1020 and 1290 A.D. (Battistini & Vérin, 1967; Burney *et al.*, 2004; Goodman & Jungers, 2013). These three sites are especially interesting as they have provided a very large number of fragments of *Aepyornis* eggshells, this accumulation being interpreted as a possible communal egg-laying site site (Battistini & Vérin, 1972; Parker Pearson *et al.*, 1995; Clarke *et al.*, 2006). No dating has been provided for Cap Sainte Marie but, on the other hand, some *Aepyornis* eggshells from Talaky have been estimated to date back to between 1020 and 1290 A.D. (Battistini *et al.*, 1963; Burney *et al.*, 2004) and those from Faux Cap to between 12980 B.C. and 530 A.D., making the latter site one of the few sites that contain remains ranging from the Pleistocene to the Holocene (Marden, 1967; Clarke *et al.*, 2006; Goodman & Jungers, 2013).

3.3.5.2. *Beraroha and Maroaloke*

The last two excavation sites in the southern part of Madagascar have not been studied in depth (Figure 3.3). All of them have provided fragments of eggshells referred to Aepyornithidae. Only the eggshells in Beraroha have been dated and seem to indicate an age of 2390±85 years B.P., that is, around 1147 B.C. (Marden, 1967; Clarke *et al.*, 2006).

3.3.6. *Outside Madagascar*

An astonishing fact that deserves to be mentioned is the discovery of two complete eggs of Aepyornithidae on the western coast of Australia where, however, no bone referred to these birds has ever been described (Long *et al.*, 1998; Tooher, 1998; Long, 2002). The hypothesis put forward to explain the presence of these eggs in Australia whereas these birds are native to Madagascar is that they may have floated between these two countries by following the marine currents of the Indian Ocean (Balanoff & Rowe, 2007).

3.4. Anatomy

Figure 3.4. *Skeleton and restoration of* Aepyornis

All Aepyornithiformes were very large birds that could reach a height of nearly 3 meters in normal standing position (Monnier, 1913) (Figure 3.4). The different genera and species are anatomically known to widely varying degrees, as we know the nearly complete skeletons of some (e.g. *A. maximus*) whereas others have merely been described based on a single bone (e.g. *A. gracilis*, which we know only from its femur). In the family Aepyornithidae, the genus *Aepyornis* is the largest and most robust, being nearly 2.70 meters tall (Clarke *et al.*, 2006), whereas among the Mullerornithidae *Mullerornis* is smaller and slenderer (Lavauden, 1931). Both genera have long legs, vestigial wings and a relatively long and robust neck (Lamberton, 1934; Clarke *et al.*, 2006) (Figure 3.4).

The anatomical study of *Aepyornis* was carried out in detail in Andrews' (1897) and Burckhardt's (1893) works, then summarized and complemented by Monnier (1913) in the light of his new taxonomy for this genus. The genus *Mullerornis* was studied first by Andrews (1904) and then followed by Lamberton's review (1934).

The head of Aepyornithiformes is relatively small compared with the rest of their body, with a straight, conical, robust and unhooked beak (Milne-Edwards & Grandidier, 1894; Monnier, 1913; Lamberton, 1934) (Figure 3.5). The cranial anatomy of all *Aepyornis* species is nearly identical and the main difference lies in their size. It can be noted that the head of *A. hildebrandti* is comparatively bigger than that of *A. maximus* and *A. medius*, even though the size of the species is smaller. The beak and mandible are also shorter, giving a head profile that is less sharp than that of the other two species (Monnier, 1913). The head of *Mullerornis* is very similar to that of *A. hildebrandti*, which is the smallest among the *Aepyornis* species (Lamberton, 1934). The diameter of the occipital condyle of *Mullerornis agilis* is significantly smaller than that of *Aepyornis*, making the facial portion of the skull delicate. Lamberton suggests that this feature is related to a diet that does not require any pecking. Some sorts of punctuations have been described on the upper part of the skull of *Aepyornis* and *Mullerornis* and interpreted as the place where a fleshy organ or the head feathers of these birds were attached (Lamberton, 1934).

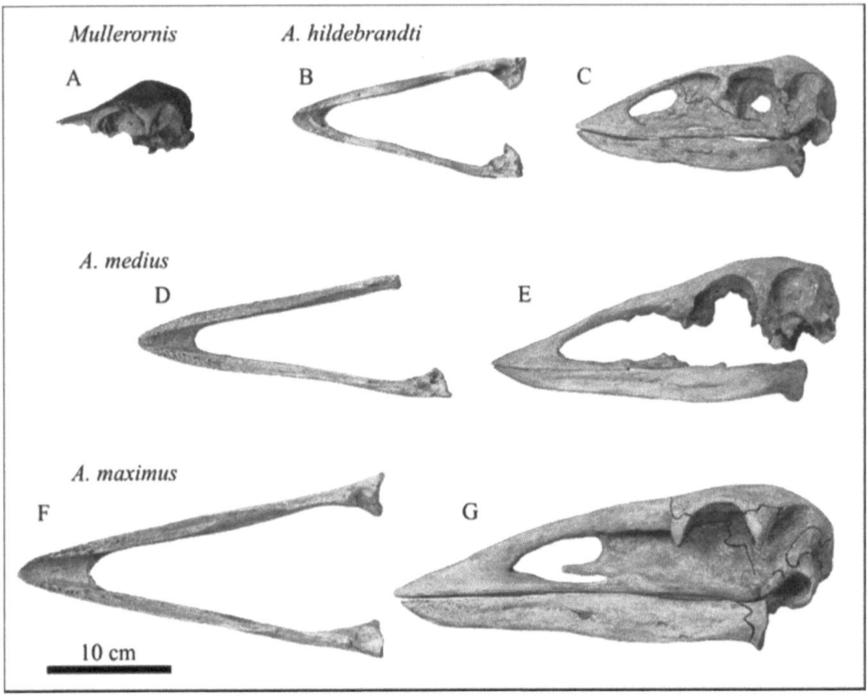

Figure 3.5. *Mandibles (B, D, F) and skulls (A, C, E, G) of Aepyornithiformes,* Mullerornis *according to Lamberton (1934), and* Aepyornis *according to Monnier (1913)*

As was the case for the cranial elements, the vertebrae of all *Aepyornis* are very similar, except for their sizes (Monnier, 1913). Andrews (1987) described the spinal column as consisting of 28 free vertebrae (20 cervical vertebrae and 8 thoracic vertebrae) besides the 20 pelvic vertebrae fused into the synsacrum. However, Monnier (1913) modified these values based on extant ratites and suggested the existence of 24 free vertebrae, among which 17 were cervical and 7 thoracic, as is the case for emus, or 16 and 8 as is the case for cassowaries. A specific feature of *Aepyornis* as compared with other ratites is that the 24th free vertebra presents a massive vertebral body that is longitudinally shortened with spinous processes that are very developed in the transversal plane (Andrews, 1897; Monnier, 1913). The vertebrae of *Mullerornis* are very similar to those of *Aepyornis*, the only difference being that they are smaller and more laterally compressed (Lamberton, 1934).

The pelvis is also anatomically very similar in all *Aepyornis* and only its size varies (Burckhardt, 1893; Monnier, 1913). All the pelvic elements (vertebrae, ilium and pubis) are firmly fused and it is hard to discern their boundaries, so that the pelvis seems very robust. Monnier (1913) points out that the shape of the pelvis is wider in females than in males, whose pelvis is longer and narrower. However, he gives no information about the method he employs to distinguish between the two sexes of *Aepyornis*. The pelvis of *Mullerornis* is narrow (three times longer than wide) and its rear part is relatively thin (Andrews, 1904; Lamberton, 1934). Lamberton (1934) concludes that, due to the gaps between the pelvic bones, *Mullerornis* could not lay very large eggs (probably smaller than those of extant ostriches).

The bones of the hindlimbs of *Aepyornis* are very robust. The femur of all species of *Aepyornis* is very robust and short, with a very wide diaphysis (Milne-Edwards & Grandidier, 1894), and strongly pneumatic (Monnier, 1913). The femur presents a large development of the trochanter and widely unequal condyles, which indicate strong muscular attachments (Monnier, 1913). The femur of *A. gracilis* (represented only by this bone) is clearly slenderer, longer and less robust than that of the other three species (Figure 3.6). The tibiotarsus of *Aepyornis* is especially robust (Milne-Edwards & Grandidier, 1894) and presents a strong longitudinal ridge for the attachment of a powerful muscle from the femur (Monnier, 1913). The tarsometatarsus of this bird is very short, flattened and wide (Monnier, 1913). Among the tarsometatarsi studied by Monnier, none shows evidence of a fourth digit (or spur). Finally, the terminal phalanges are mostly unhooked and broad. The bones of the hindlimbs of *Mullerornis* are very similar to those of *Aepyornis* (Lamberton, 1934). The femur is more slender and its head is relatively smaller; similarly, the tibiotarsus is thinner. Finally, the tarsometatarsus is short, strong and widened at the distal end (Milne-Edwards & Grandidier, 1894). The tibiotarsi of *M. betsilei* and *M. rudis* are of similar lengths, but the tibiotarsus of *M. betsilei* is more robust and its foot is not widened (Figure 3.6). The phalanges of *Mullerornis* were quite similar to those of *Aepyornis*. However, the terminal phalanges were longer and sharper (Andrews, 1904; Lamberton, 1934).

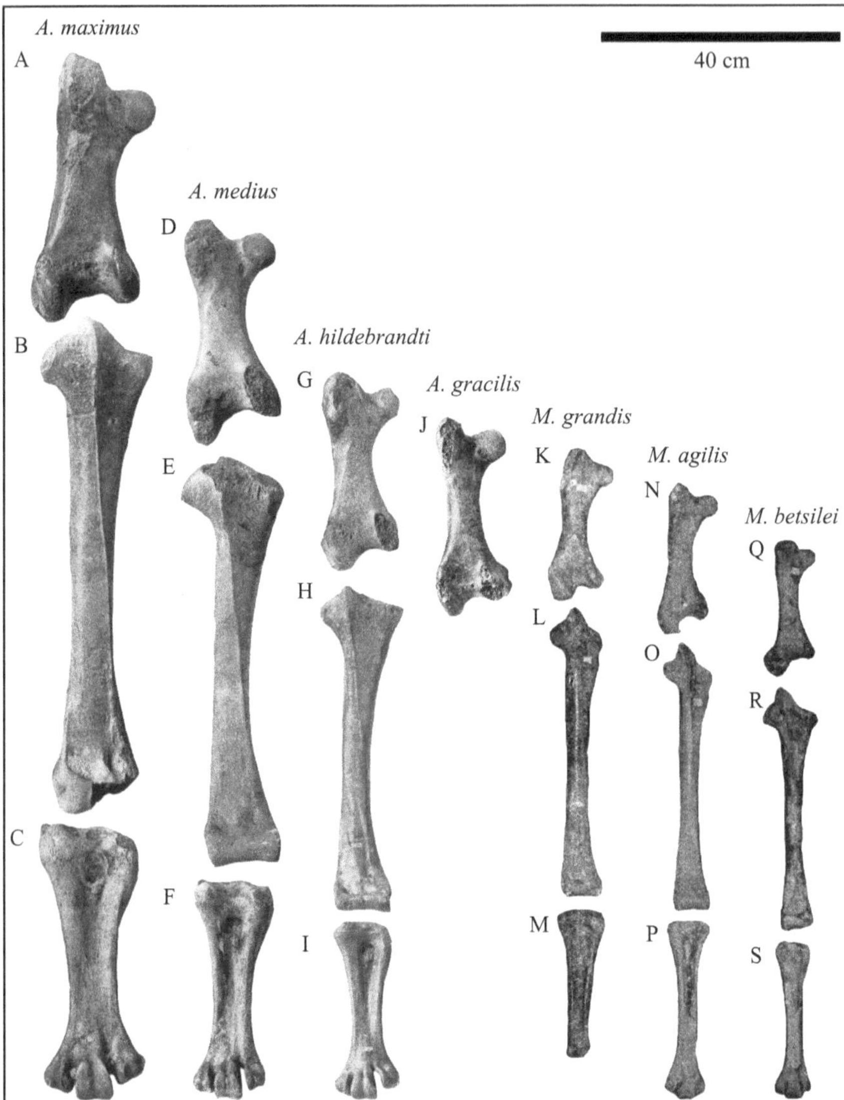

Figure 3.6. *Differences in the hindlimbs bones of Aepyornithiformes. A, D, G, J, K, N, Q: femora; B, E, H, L, O, M: tibiotarsi; C, F, I, M, P, S: tarsometatarsi. Images A–J are from Monnier (1913) and images K–S from Lamberton (1934)*

3.5. Paleophysiology and paleoecology

3.5.1. *Diet*

The first to mention the diet of Aepyornithiformes was Geoffroy Saint-Hilaire (1851), who concluded his description of the first specimens of *Aepyornis maximus* by saying that these birds "had neither talons nor wings fit for flying, and that they must have peacefully fed on plant material". Afterwards, Lavauden (1931) thought that these birds must have been forest dwellers based on the very large bones of their legs, which must have allowed them to scratch the surface to extract larvae, worms and roots. It was also suggested that these birds fed on nearshore and lagoonal shellfish, which would explain the fact that they can be found in abundance in coastal areas (Lamberton, 1934). In this case, their very flexible legs would have allowed them to hold shells on the ground while they pecked at them with their beaks. However, the anatomy of their beak and lower jaw does not support this hypothesis and seems to be in line with an herbivorous diet, as is the case for all present-day ratites.

It was not until 2006 that the studies carried out by Clarke *et al.* (2006) and then Tovondrafale *et al.* (2014) provided more evidence based on the study of isotopes. The carbon isotope analysis of the eggshells attributed to *Aepyornis* does not only demonstrate that these birds were actually herbivorous, but provides additional evidence. Thanks to the study of the $\delta^{13}C$ of the eggshells, we can estimate the proportions of C3 plants (trees, shrubs, grass, etc.) and CAM plants (plants such as succulents that have adapted to hot and dry environments) consumed by these birds (the plants present in the Madagascan environment of this period). The $\delta^{13}C$ of the eggshells of Aepyornithidae is approximately 25.2‰, that of C3 plants present at these sites is around 13.5‰ and that of CAM plants is approximately 26.4‰ (values measured at the sites where fossil eggshells have been sampled) (Clarke *et al.*, 2006). Based on these values, Clarke *et al.* (2006) have been able to show that these birds consumed more C3 plants (around 91%) than CAM plants (around 9%).

More recently, Tovondrafale *et al.* (2014) have carried out the same type of measurement on other sites in the southern part of Madagascar, obtaining a $\delta^{13}C$ value of 24.2‰ for the eggshells and a $\delta^{13}C$ value of around 12‰ for the C3 plants present in this environment, which rises to around 22‰ for CAM plants. However, they have not taken into consideration a possible mixture of food supplies, making their interpretation invalid. By reconsidering their values and calculating again the proportions of food, we also obtain as a result that *Aepyornis* preferred feeding on C3 plants (88%) rather than CAM plants (12%). For all the sites studied, it has then been shown that these birds preferred feeding on the leaves of trees (C3) rather than grazing on grass on the ground and the abundant local flora of CAM plants, in line with the anatomy of Aepyornithidae, which had a long neck and long legs in order to easily reach the leaves of trees (Clarke *et al.*, 2006). It is also interesting to note that the study carried out by Tovondrafale *et al.* (2014) was centered on specimens from both the Pleistocene and the Holocene. These authors conclude that there have been no changes in the dietary habits of *Aepyornis* over time. However, if we reconsider their data, it seems that there has actually been an evolution in the diet of *Aepyornis* between the Pleistocene and the Holocene, passing from a diet based 93% on C3 plants in the Pleistocene to a diet including less C3 plants (84% of C3 plants and 17% of CAM plants) in the Holocene. This evolution of their diet in the direction of more CAM plants in the Holocene may be associated with the aridification of the environment during this period in the southern part of Madagascar (Clarke *et al.*, 2006; Tovondrafale *et al.*, 2014), which may have favored CAM plants more adapted to this type of environment. These climate changes may have thus increased the proportion of CAM plants in the living environment of *Aepyornis*, as Clarke *et al.* (2006) have suggested, and thus affect the proportions of plants on which these birds fed.

The anatomy of *Mullerornis* shows that they were also herbivorous. The only $\delta^{13}C$ values (3 values) for eggshells referred to this genus have been measured by Tovondrafale *et al.* (2014), who obtained an average value of around 22‰, suggesting that the diet of these birds included around 70% of C3 plants and approximately 30% of CAM plants. This tends to show that their diet was slightly different from that of *Aepyornis*, which may explain how these two genera shared the ecological niches of their living environment. Nonetheless, this conclusion is only based on three $\delta^{13}C$ measurements, so that more analyses should be carried out before confirming this trend.

The herbivorous diet of Aepyornithiformes has potentially affected the defense strategies of Madagascan woody plants, as Bond and Silander suggested (2007). In order to protect themselves against these browsing animals, plants have developed forms of defense such as small leaves that reduce the leaf area lost to browsing, the presence of thin branches that are hard to bite and have high tensile strength so as not to break easily when pulled, or elastic branches that work like springs once a bird has left the branch, increasing handling time during browsing.

3.5.2. *Body mass*

The first body mass estimate for Aepyornithiformes was put forward by Lavauden (1931) and corresponded to 500 kg. Subsequently, Amadon (1947) estimated the weight of *Aepyornis maximus* to be around 400 kg. However, no actual method for estimating body mass was used in these two cases. At a later stage, several body mass estimates obtained using different methods were put forward for Aepyornithidae (Table 3.1). Thus, *Aepyornis maximus*, which is the largest bird in this family, has an estimated average body mass that ranges between 310 kg and 747 kg. Only the estimate made by Nguyen *et al.* (2010) is significantly smaller, corresponding to a value of 243 kg. Nonetheless, it is quite likely that this specimen does not belong to *A. maximus* but to one of the other smaller species. As taxonomy is very chaotic, this type of problem is quite common. Similarly, it is certainly possible that part of the body mass estimates put forward for *A. maximus* actually correspond to other species of *Aepyornis*.

The only average body mass estimates put forward for *Mullerornis* range from 74 to 92 kg, which is significantly less than what has been calculated for *Aepyornis*, thus supporting the idea of a smaller and more slender bird. However, in the case of *Mullerornis* too, estimates have been made based on specimens of *M. agilis*, but since taxonomy is not clearly defined, other species or a mixture of several species may have been used to carry out these calculations.

Despite taxonomic uncertainties, *Aepyornis* was significantly heavier and larger than *Mullerornis*, with a mean body mass of 500 kg and 90 kg, respectively.

Methods	Campbell & Marcus (1992) (femur) (kg)	Campbell & Marcus (1992) (tibiotarsus) (kg)	Anderson *et al.* (1985) (femur) (kg)	Hurlburt (1999) (GDI model) (kg)	Other or unspecified (kg)
Aepyornis maximus.	542 (333–725) [2–3] 728–747 [7] 635 (627–644) [7] 528 (342–714) [9]	243 [8] 361 (243–478) [9]	695-716 [3] 377 (216–661) [4]	440–516 [3]	438 [1] 418–457 [5] 310 [6]
Aepyornis sp.	515 (220–776) [7]	300 (165–537) [12]			500 [8]
Mullerornis agilis	92 (75–118) [7]	74 (60–82) [7]			

Table 3.1. *Compilation of estimated body mass values for Aepyornithiformes from the literature. Average value (minimum value – maximum value); [bibliographical references]: 1 – Alvarenga & Höfling (2003), 2 – Campbell & Marcus (1992), 3 – Murray & Vickers-Rich (2004), 4 – Nguyen* et al. *(2010), 5 –Amadon (1947), 6 - Worthy & Holdaway (2002), 7 – Angst (2014), 8 – Lavauden (1931) 9 – Handley* et al. *(2016)*

3.5.3. *Locomotion*

One of the first hypotheses concerning the locomotion of Aepyornithiformes resulted from the phylogenetic hypothesis put forward by Bianconi (1870). Taking into consideration that these birds must be giant birds of prey similar to condors, Bianconi suggested that the tibiotarsus was attached to a developed extensor muscle and an atrophied flexor muscle to make jumping easier when taking off (Bianconi, 1870). Later, Monnier (1913) suggested a "flightless bird" type of locomotion, relying on the more or less legendary description of a "large beast whose two legs were situated at the center of its stomach and which left, as its name tells us, only one mark on the ground", which he thought could only correspond to a flightless bird. This idea was taken up again later on by Lamberton (1934), but this author was already emphasizing that the tarsometatarsus of these birds was particularly short, unlike that of other present-day ratites such as ostriches.

More recently, it has been possible to reassess the locomotion of Aepyornithiformes thanks to the method used by Angst *et al.* (2016). Whether *Aepyornis* or *Mullerornis*, all the birds included in this order had a relatively slow and graviportal locomotion type (Angst *et al.*, 2016; Angst &

Chinsamy, 2016, in press). However, it should be pointed out that *Aepyornis* has a more robust tarsometatarsus than *Mullerornis*, with proportions of 5.5 and 10.8, respectively (Figure 3.7). Consequently, the former was more graviportal than the latter, which was probably due to the significant difference in body mass between these two genera (Angst & Chinsamy, in press).

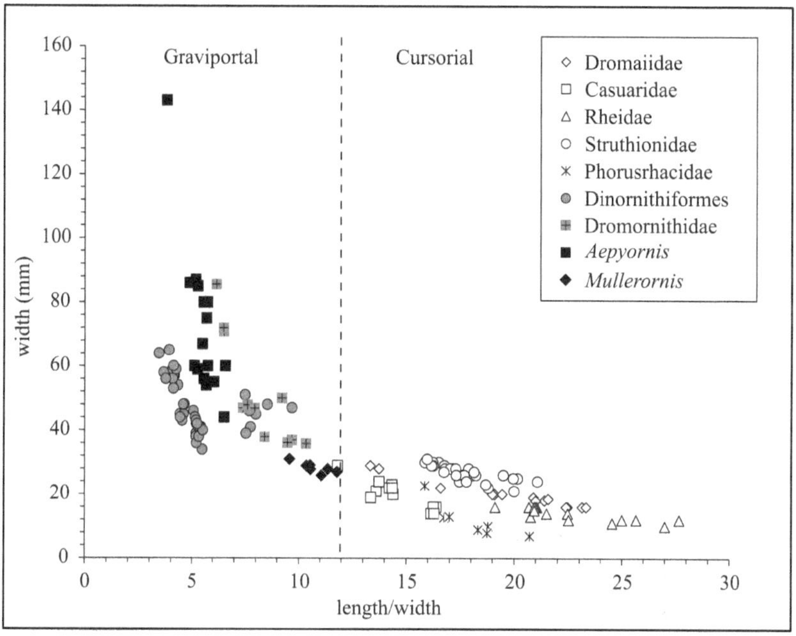

Figure 3.7. *Estimate of the type of locomotion of Aepyornithiformes based on the method used by Angst* et al. *(2016)*

3.5.4. *Reproduction*

It is possible to study the reproduction of Aepyornithiformes thanks both to the very abundant eggshell fragments available and to complete eggs and, more specifically, the skeleton of an embryo contained in one of these eggs.

The eggs of Aepyornithiformes were the first fossils attributed to these large birds that were found and described in detail (Geoffroy Saint-Hilaire, 1851; Lavauden, 1931). These eggs generally measure 30 cm in length (ranging from 26 cm to 40 cm) with an average width of 22 cm (ranging from 19 cm to 25 cm) (Geoffroy Saint-Hilaire, 1851; Lavauden, 1931; Henrici, 1957; Mlíkovský, 2003; Balanoff & Rowe, 2007; Angst *et al.*, 2014b) (Figure 3.8), corresponding to an average volume of around 7700 cm^3 (i.e. 7.7 liters; ranging from about 5600 cm^3 to about 13000 cm^3) and an average weight of 8 kg (ranging from around 6 kg to around 14 kg) (Amadon, 1947; Balanoff & Rowe, 2007; Angst *et al.*, 2014b). A statistical study carried out by Mlíkovský (2003) based on the measurements of the length and width of 43 entire eggs has demonstrated that all these eggs may be assigned to a single species, *Aepyornis maximus*. The grain of the surface of these eggs varies. Some eggs have a polished and occasionally even shimmering surface, which may be explained by an abrasion of the surface caused by the action of sand-carrying wind, whereas others present a coarser grain similar to that observed on the eggs of present-day *Dromaius* (Lavauden, 1931).

Figure 3.8. *Eggs of* Aepyornis *(on the left, photo E. Buffetaut) compared with a hen's egg (on the right, photo D. Angst). For a color version of the figure, see www.iste.co.uk/angst/birds.zip*

3.5.4.1. *The specific attribution of eggs*

In order to use these eggs to study the reproduction of these birds, it is necessary to assign them to Aepyornithiformes. They have been assigned to them by adopting a mainly geographical approach. No other ratite big enough to have laid eggs of this size is known to have lived on the island of Madagascar (Balanoff & Rowe, 2007). Among the known eggshell fragments, some are thicker than others. Based on this observation, the thicker shells (2 – 3.9 mm) were assigned to the genus *Aepyornis*, whereas the thinner ones (0.5 – 1.7 mm) were associated with *Mullerornis* (Geoffroy Saint-Hilaire, 1851; Arnadon, 1947; Burney *et al.*, 1997; Clarke *et al.*, 2008; Tovondrafale *et al.*, 2014).

3.5.4.2. *Egg-laying sites and periods*

Clarke *et al.* (2006) suggest that *Aepyornis* laid eggs in wintertime based on the breeding season of extant ratites (Crome, 1975; Bertram, 1992; Coddington & Cockburn, 1995). However, they had no other element available to test this hypothesis.

The significant accumulation of eggshell fragments on the beaches of the southern and south-western part of Madagascar (Rowley, 1867; Grandidier, 1868; Monnier, 1913) has been thought to correspond to possible egg-laying areas for Aepyornithiformes (Last, 1894). This accumulation may be due to several factors. It may be either the result of shared egg-laying areas or a blend of shells of different ages. In the absence of more extensive dating for all the shells of a given site, it remains difficult to choose between the two hypotheses.

Finally, thanks to oxygen and carbon isotopic analyses carried out on the eggshells, it has been possible to obtain information about the diet and drinking water of *Aepyornis* during the breeding season. Thus, thanks to oxygen isotopes, Clarke *et al.* (2006) have been able to show that these birds drank the water of ponds supplied by groundwater. Moreover, they have demonstrated that the diet of *Aepyornis* included mostly leavcs of C3 plants (see above). The flora of the environment in which the eggshells have been found is dominated by CAM succulents. The difference between the flora of the environment and the plants eaten by these birds may be explained with two hypotheses: either *Aepyornis* selected the plants they fed on by consuming mainly the C3 plants present in their nesting sites or they nested in thickets of CAM plants but fed in the surrounding forests that included mostly C3 trees. In the former case, the birds nested and fed in the same place while also choosing their food, whereas in the second case they did not feed in the same place where they laid their eggs, moving towards areas richer in C3 plants to feed (Clarke *et al.*, 2006).

3.5.4.3. *Embryos*

The first embryonic skeleton inside an *Aepyornis* egg was observed by Wetmore (1967) thanks to X-ray images. Unfortunately, the images did not make it possible to carry out a specific study of this embryo. The only study currently available is the one carried out by Balanoff and Rowe (2007) on an *Aepyornis* egg containing an embryo, which they have scanned through high-resolution X-ray tomography. This embryo represents the most complete and best-preserved known specimen of *Aepyornis*. However, as it is only at an embryonic stage, bones have not finished developing and suturing, occasionally making it hard to compare the embryo with adult specimens. Nonetheless, the embryonic anatomical features have also allowed these authors to assess the ontogenetic maturity (or embryonic stage) of this specimen by comparing it with embryos of present-day ratites. We know, based on the number of days of incubation of ratites, when the mineralization of each bone starts or when bones begin fusing.

There are several ways of assessing the embryonic stage, but Webb's way (1957), which links the age of the embryo in days to a percentage of incubation, is the easiest to interpret. This conversion is necessary to be able to compare the embryonic development of different birds, as the incubation period varies from species to species. Thus, in order to compare the embryonic development of *Aepyornis* with other ratites, we need to be able to tell whether the incubation of a specimen, for example, has reached 30% of its term or has been going on for three days. To carry out this standardization, the number of days after which an event happens is divided by the length of the incubation period (in number of days). Therefore, if the incubation period is of 50 days and an event happens on the fifteenth day, it will be possible to tell that we have reached 30% of the incubation period. However, we can calculate this only if we know the total length of the incubation period of the bird. For *Aepyornis*, the duration of the embryonic development has been estimated at 85–90 days thanks to different equations (Needham, 1931; Worth, 1940; Ar *et al.*, 1974). Based on all these data, Balanoff and Rowe (2007) have been able to determine that the embryo contained in the egg they have studied had reached around 80–90% of its incubation period when it died.

Studying this embryo has also made it possible to determine that the major factor in the evolution of *Aepyornis* towards such a large size plays a role very early in the ontogenesis, before the eggs hatch, with a very significant growth rate for hindlimbs since the embryonic stages. However, forelimbs are much reduced, suggesting strong variations in the growth rate in relation to the different parts of the skeleton. Thus, compared with other ratites, the growth of hindlimbs accelerates (peramorphosis) and the development of forelimbs slows down (paedomorphosis) (Balanoff & Rowe, 2007).

3.6. Paleoenvironment

Aepyornithiformes lived in an environment of undergrowth or open woodland forests bordering more or less marshy plains (Milne-Edwards & Grandidier, 1894; Lamberton, 1934). This is backed up by palynological studies (Burney, 1993), which indicate that there were more arborescent species in Madagascan forests before 1000 A.D., that is, when Aepyornithiformes lived on the island. Thus, the coastlines where most fossil sites have been found may have been covered by more trees and shrubs than today, thus allowing Aepyornithiformes to feed more easily (Clarke *et al.*, 2006). This type of environment is also compatible with the $\delta^{13}C$ values measured on the eggshells of *Aepyornis*, which indicate that these birds fed mainly on tree or shrub leaves (Clarke *et al.*, 2006).

The fossil sites that contained the remains of Aepyornithidae also indicate that those environments were moderately humid and corresponded to bodies of water such as lakes or freshwater marshes where these birds came to drink (Last, 1894; Monnier, 1913; Clarke *et al.*, 2006; Balanoff & Rowe, 2007). This is supported by the fauna associated with Aepyornithiformes in the different sites and including freshwater aquatic animals such as pigmy hippopotamuses, crocodiles or turtles (Milne-Edwards & Grandidier, 1894; Clarke *et al.*, 2006). However, these bodies of water were situated in a globally semi-arid environment, as shown by the $\delta^{18}O$ values measured on the eggshells of *Aepyornis* (Clarke *et al.*, 2006) with an aridification of the environment between the Pleistocene and the Holocene, as is demonstrated by the measurements of the $\delta^{15}N$ values (Tovondrafale *et al.*, 2014).

3.7. Extinction

3.7.1. *The extinction date of Aepyornithiformes*

The extinction date of these large birds is not known with precision as of now. The account of the French governor of Madagascar, Etienne de Flacourt, suggests that some Aepyornithiformes were still alive around 1650, even though there were very few of them (Lavauden, 1931). On the other hand, Grandidier and Milne-Edwards stated in 1869 that there was no longer any individual of Aepyornithiformes alive, even though their complete extinction had taken place rather recently (Grandidier, 1867a, 1867b; Milne-Edwards, 1867, 1873, 1872; Monnier, 1913). Apart from these accounts, whose reliability is hard to test, the carbon-14 dating of the most recent remains of *Aepyornis* corresponds to 1290 A.D. (Marden, 1967; Burney *et al.*, 2004) and was carried out on a fragment of eggshell found at Fort Dauphin. This value shows that these birds became extinct during the end of the late Holocene and, therefore, lived at the same time as humans in Madagascar from around 350 B.C., which represents the estimated date of the arrival of the first humans on the island, during the Indonesian Iron Age (Deschamps, 1961; Mahé & Sourdat, 1972; Burney *et al.*, 2004; Gommery *et al.*, 2011) and potentially as late as the 17th Century in some more isolated areas (Lavauden, 1931).

3.7.2. *Possible causes for the extinction of Aepyornithiformes*

Several causes have been considered to be behind the extinction of Aepyornithiformes. Part of these hypotheses involve environmental changes independent of the arrival and presence of human beings in Madagascar, whereas others are directly linked to human activities.

3.7.2.1. *Environmental changes*

Between the end of the Pleistocene and the beginning of the Holocene (before the arrival of the first humans), significant climate changes took place and altered the flora of the island (Burney *et al.*, 2004). The study of pollens (Burney, 1993; Clarke *et al.*, 2006) and the variations in $\delta^{15}N$ deriving from the organic phase of the eggshells of Aepyornithiformes (Tovondrafale *et al.*, 2014) show that a phase of aridification associated with the drying up of the wetlands in the south-western and southern parts of the island took place in Madagascar from the end of the Pleistocene and at the beginning of the Holocene (Mahé & Sourdat, 1972; Burney, 1997; Rakotozafy & Goodman, 2005). These environmental changes probably reduced the abundance of C3 plants by favoring the development of CAM plants that were more adapted to the dry climate (Clarke *et al.*, 2006). As Aepyornithiformes fed mainly on C3 plants (between 70 and 93%), this evolution of the plant cover must necessarily have had an impact on these birds. Tovondrafale *et al.* (2014), and Goodman and Jungers (2013) go as far as saying that these environmental changes must have had a more significant impact than human activities.

3.7.2.2. *Human activities*

Human activities represent the second factor that must have played a direct or indirect role in the extinction of Aepyornithiformes. The most evident causes are clearly linked to the hunting of these birds in which the people inhabiting the island engaged, as shown by the remains of sharp tools found on bones (Lavauden, 1931). Hunting must also have been facilitated by the concentration of these birds around bodies of water during nesting as well as the concentration of eggs in nesting areas that were also close to these bodies of water, making it easier to gather eggs and weakening this group even more (Clarke *et al.*, 2006). In numerous archeological sites, significant quantities of eggshells of Aepyornithiformes can be found, suggesting that they were commonly consumed (Gommery *et al.*, 2011).

Furthermore, human activities may have had indirect effects. The first settlers brought to the island cattle that must have started competing with Aepyornithiformes for access to part of the flora (Clarke *et al.*, 2006). Limited access to food supplies for these birds must have been amplified by the introduction and development of slash-and-burn agriculture in Madagascar, which made it possible to clear large land surfaces, as has been shown by the sudden increase in the quantity of coal present in the palynological records during the Holocene (Burney, 1997; Clarke *et al.*, 2006). At the same time, deforestation also increased, thus dividing up the range of these birds and therefore affecting the evolution of the populations (Lavauden, 1931). Finally, the arrival of the first settlers in Madagascar may have introduced new viral and contagious diseases that probably exterminated part of the indigenous fauna, which had not been immunized to them (Clarke *et al.*, 2006).

The exact causes behind the extinction of Aepyornithiformes are not yet completely understood and remain a topic of debate. The extinction of these birds may very probably be the result of a combination of causes (Crowley, 2010), at least some of which are associated with the arrival of humans in Madagascar. The relatively long cohabitation of these birds with humans on the island (between 350 B.C. and 1290 A.D., and even longer if Flacourt's account is true) probably accentuated the impact of humans on the populations of Aepyornithiformes until their extinction. Nonetheless, this long cohabitation also shows that the arrival of humans on the island did not cause their immediate extinction and that the disappearance of Aepyornithiformes from Madagascar is probably due to multiple factors (Gommery *et al.*, 2011).

4

Dromornithidae

4.1. Historical background

Dromornithids were first mentioned in aboriginal oral tales. Specifically, the members of the Tjapwrong tribe in the western region of Victoria referred to "mihirung parinhmal", namely giant emus that must have lived long ago, when the volcanoes in the western part of Victoria were still active. The most recent volcanic flows in this region date back to 1400 B.P. (i.e. 550 A.D.) (Gill, 1972; Blackburn *et al.*, 1982), but it is difficult to decide whether these traditions are based on an actual cohabitation with these birds while they were still alive or if they concern bones. Afterwards, cave paintings representing a giant bird of the same size as emus and possibly depicting *Genyornis newtoni* were discovered in caves in the northern part of Queensland (Tresize, 1971; Vickers-Rich *et al.*, 1997) and in the western region of Arnhem Land, in the north of the Northern Territories (Gunn *et al.*, 2011). Once again, it is difficult to determine whether these drawings have been inspired by an actual bird or come from the imagination of the artist and, in the first case, if artists were inspired by an actual meeting with the bird or by descriptions passed on orally.

The first specimen (a femur) that clearly belongs to Dromornithidae was discovered in 1839 in the Wellington Valley (in present-day New South Wales), during an expedition conducted by Major T. L. Mitchell (Mitchell, 1838), and sent to Richard Owen to be studied. Unfortunately, this specimen was destroyed in the bombing of London during the Second World War. The first convincing proof that Dromornithidae were unique birds was provided by the discovery of a complete skeleton in 1892 in lake sediments in the north-eastern part of Southern Australia. This finding was followed by excavations that led to the discovery of a large amount of additional material that was studied by Newton a year later, allowing him to confirm the presence of a large fossil bird the size of an ostrich in Australia.

Since the end of the 19th Century, a large number of expeditions were conducted, enabling the discovery of a larger array of specimens. During an expedition led by Reverend J.E.T. Woods in 1866, bones of Dromornithidae that presented traces of tools typical of local aboriginals were discovered, so that the cohabitation between men and these birds could be proved (Woods, 1866). Unfortunately, these bones now cannot be found. A large number of additional fossils were discovered later on and studied, among others, by Sir Richard Owen (1874, 1879a-b), Etheridge (1889), Stirling and Zietz (1900) and Rich (1979). In 1962, footprints were discovered in sediments dating back to the middle of the Tertiary and assigned to Dromornithidae by Rich and Green (1974). Finally, at the end of the 19th Century, fragments of eggs were discovered and referred to Dromornithidae, more specifically to *Genyornis newtoni*, by Miller *et al.* (1999).

4.2. Taxonomy, systematics and classification

4.2.1. *Taxonomy*

The family Dromornithidae was established by Fürbringer (1888) on the basis of inadequate material and it has been recently redefined by Murray and Vickers-Rich (2004), who have proposed a new diagnosis of this family, which includes large flightless birds reminiscent of geese and belonging to the order Anseriformes.

Eight species have been described in this family (Murray & Vickers-Rich, 2004), including two species (*Ilbandornis? lawsoni* and *?Bullockornis* sp.) whose genus had been incorrectly identified until recently. The different species were mainly differentiated thanks to size differences and some anatomical features observable on the bones of their hindlimbs (cranial remains were in most cases too limited and poorly preserved). Woods described the first Dromornithidae on the basis of bones discovered in a marsh close to Penola (Woods, 1866). He proposed the name *Dromaius australis*, but unfortunately he did not illustrate his type specimens and provided no precise description of the fossils, making this species a *nomen* *dubium*. This species became valid only after Owen's work (1872), based on a femur discovered in Peak Downs which he attributed to *Dromornis australis*. Afterwards, the species *Genyornis newtoni* was described on the basis of a femur and then a subcomplete skeleton discovered in 1892 (Stirling & Zietz, 1896). More than 80 years elapsed until Patricia Vickers-Rich reviewed the available evidence on Dromornithidae (Rich, 1979), put forward really complete diagnoses for the two aforementioned species, and especially described six new species included in four genera: *Barawertornis tedfordi*, *?Bullockornis* sp., *Bullockornis planei*, *Ilbandornis? lawsoni*, *Ilbandornis woodburnei* and *Dromornis stirtoni*. For several decades, excavations continued and led to the discovery of a large number of new specimens (Murray & Vickers-Rich, 2004). In 2010, a new study focused on all the dromornithid material and put forward a new classification based on a large number of new synapomorphies (Nguyen *et al*., 2010). These authors suggested then that *?Bullockornis*. sp. should be synonymized with *Ilbandornis*. sp., and that *Ilbandornis? lawsoni* should be replaced by *Genyornis lawsoni* and *Bullockornis planei* by *Dromornis planei*, thus removing the genus *Bullockornis*. It is interesting to note that the first species of Dromornithidae, *Dromornis australis* (Owen, 1874), was described on the basis of its femur, explaining why the taxonomy of Dromornithidae has principally developed in relation to the morphology of this bone (Worthy & Yates, 2015). However, for *Ilbandornis*, the tarsometatarsus was chosen as type specimen, raising a certain number of comparative issues (Rich, 1979). This situation was solved by the work carried out by Worthy and Yates (2015), who were the first to reconsider all the specimens of *Ilbandornis* by including detailed comparisons with all the femora belonging to the other Dromornithidae, so as to homogenize the totality of comparative elements. This study, therefore, allowed them to confirm most of the relationships suggested by Nguyen *et al*. (2010) except for the reattribution of the fossils of *Ilbandornis? lawsoni*, which were reconfirmed as belonging to *Ilbandornis lawsoni*. Finally, the very last species of Dromornithidae was described in 2016 by Worthy *et al*. (2016) based on cranial and postcranial material discovered in sites at Riversleigh which date back to the Oligocene–Miocene. This new species, *Dromornis murrayi*, thus represents the most ancient and smallest species belonging to the genus *Dromornis*.

4.2.2. *Systematics*

Different types of systematics have been suggested for Dromornithidae since the 19th Century (Fürbringer, 1888; Rich, 1979; Murray & Vickers-Rich, 2004; Nguyen *et al.*, 2010; Worthy & Yates, 2015; Worthy *et al.*, 2016). The one put forward here constitutes a synthesis of the latest studies about this group. Dromornithidae include nine species belonging to four genera (Worthy & Yates, 2015).

Class Aves Linnaeus, 1758

Super-order Neognathae Pycraft, 1900

Order Anseriformes Wagler, 1831

Suborder Anhimae Wetmore & Miller, 1926

Family Dromornithidae Fürbringer, 1888

Genus *Genyornis* Stirling & Zietz, 1896

– *Genyornis newtoni* Stirling & Zietz, 1896

Stratigraphic occurrence: Pleistocene of Australia (Southern Australia, Western Australia, New South Wales, Queensland)

Genus *Barawertornis* Rich, 1979

– *Barawertornis tedfordi* Rich, 1979

Stratigraphic occurrence: Lower and Middle Miocene of Australia (Queensland)

Genus *Ilbandornis* Rich, 1979

– *Ilbandornis* sp. Nguyen *et al.*, 2010

Stratigraphic occurrence: Middle Miocene of Australia (Northern Territory)

– *Ilbandornis lawsoni* Rich, 1979

Stratigraphic occurrence: Upper Miocene of Australia (Northern Territory)

– *Ilbandornis woodburnei* Rich, 1979

Stratigraphic occurrence: Upper Miocene of Australia (Northern Territory)

Genus *Dromornis* Owen, 1872

– *Dromornis australis* Owen, 1872

Stratigraphic occurrence: Upper Pliocene of Australia (Queensland)

– *Dromornis stirtoni* Rich, 1979

Stratigraphic occurrence: Upper Miocene of Australia (Northern Territory)

– *Dromornis planei* (Rich, 1979)

Stratigraphic occurrence: Middle Miocene of Australia (Northern Territory)

– *Dromornis murrayi* Worthy *et al.*, 2016

Stratigraphic occurrence: Upper Oligocene–Lower Miocene of Australia (Queensland)

4.2.3. *Classification*

The first classification of Dromornithidae was put forward by Richard Owen (1872), who classified *Dromornis australis* among paleognathous ratites (Ostriches, Rheas). Later, Stirling and Zietz (1896) attempted to improve this assignment by pointing out that these birds belong to Australian Casuariiformes (Emus, Cassowaries). It took a century for the phylogenetic position of Dromornithidae to be studied again, when it was shown that they were not paleognathous birds but neognaths and, more specifically, basal Anseriformes (Ducks, Geese) (Murray & Megirian, 1998), and that their similarity with paleognaths was a mere evolutionary convergence. Dromornithidae are anatomically intermediate between two extant groups of Anseriformes, namely Anhimidae and Anseranatidae, with some features that make them more similar to the former group. The most recent phylogeny that includes Dromornithidae is that put forward by Mayr (2011), who classifies these birds as the sister-group of Galliformes, even though the author himself acknowledges that his study may contain inadequate characters (Figure 4.1). The exact phylogenetic position of Dromornithidae, therefore, is still debated (Figure 4.1), and so are the phylogenetic relations between the different genera and species of Dromornithidae (Murray & Vickers-Rich, 2004; Nguyen *et al.*, 2010).

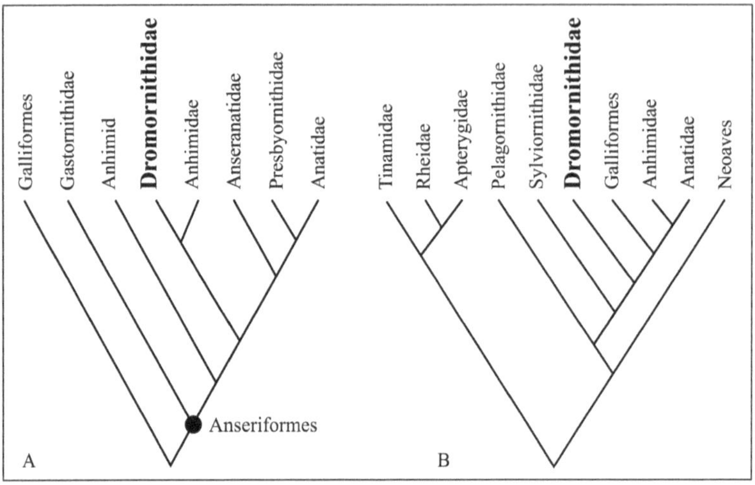

Figure 4.1. *Phylogenetic position of Dromornithidae according to Murray & Vickers-Rich (2004, A) or Mayr (2011, B)*

4.3. Stratigraphic and geographical distribution

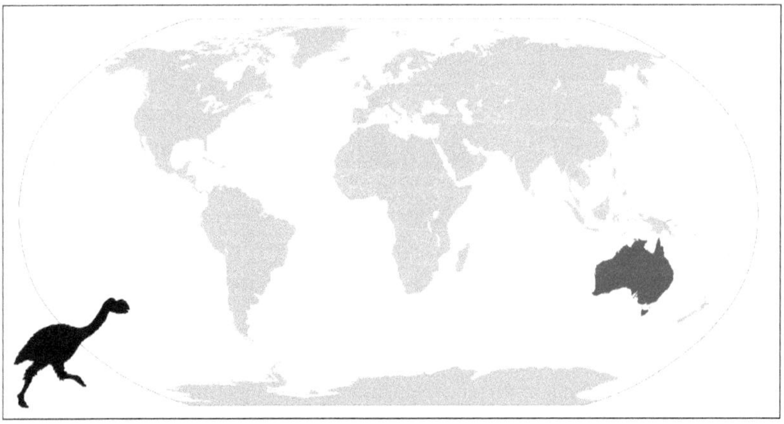

Figure 4.2. *Global geographical distribution of Dromornithidae. For a color version of the figure, see www.iste.co.uk/angst/birds.zip*

Dromornithidae are known only from Tertiary (starting from the end of the Eocene) and Quaternary sediments (until the end of the Pleistocene) in continental Australia and Tasmania (Murray & Vickers-Rich, 2004). These birds thus represent the warm-blooded vertebrates with the longest fossil record in Australia. The first occurrence attributed to Dromornithidae corresponds to a footprint discovered in Eocene sediments (around 55 Ma) in Queensland dating back to the period when Australia started separating from Antarctica. During the Tertiary, a significant diversification of these birds into eight species took place, with their final extinction occurring at the end of the Pleistocene, when only one species is known (*Genyornis newtoni*) (Figure 4.2).

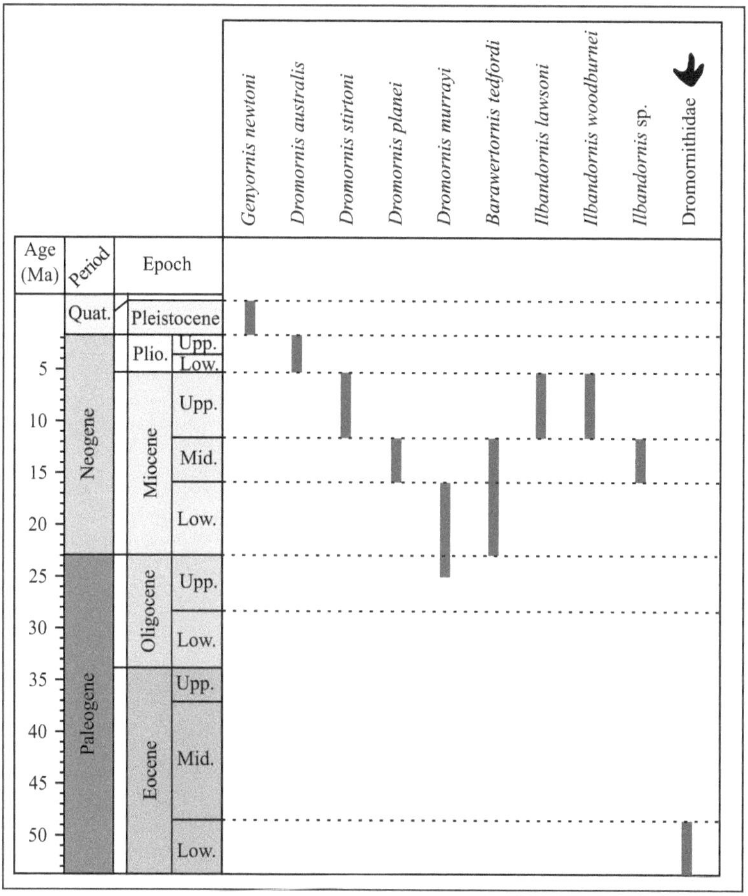

Figure 4.3. *Simplified stratigraphic distribution of Dromornithidae. : footprints. For a color version of the figure, see www.iste.co.uk/angst/birds.zip*

Dromornithidae are known from all parts of Australia, and most specimens derive from three main fossil sites: Callabonna Lake (Southern Australia, Pleistocene), Alcoota (southern part of the Northern Territory, Upper

Miocene), and Bullock Creek (northern part of the Northern Territory, Middle Miocene). Specimens of Dromornithidae represent about 25% of the fossils discovered in the last two sites (Handley *et al.*, 2016). However, fossils attributed to these birds are known from several dozen fossil sites, only the most significant of which will be described in detail here (Figure 4.3, Figure 4.4).

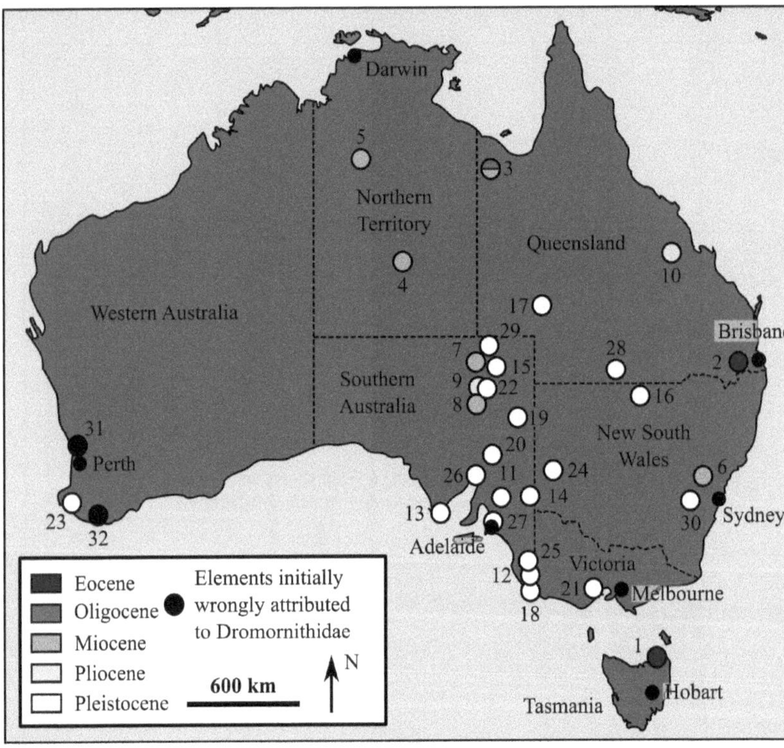

Figure 4.4. *Geographical distribution of the main fossil sites that have yielded fossils attributed to Dromornithidae. Eocene sites: 1 – Old Endurance Tin Mine, 2 – Redbank Plains; Oligocene site: 3 – Riversleigh Station; Miocene sites: 4 – Alcoota Local Fauna and Ongeva Local Fauna, 5 – Bullock Creek, 6 – Canadian Deep Lead Mine, 7 – Leaf Locality, 8 – Snake Dam Locality; Pliocene sites: 9 – Lawson Quarry, 10 – Peak Downs; Pleistocene sites: 11 – Baldina Creek, 12 – Big Cave, 13 – Brothers Island, 14 – Confluence Darling and Murray Rivers, 15 – Cooper Creek and Pitikanta Lake, 16 – Cuddie Springs, 17 – Diamantina River, 18 – Gambier Range, 19 – Lake Callabonna, 20 – Lake Frome and Hookina Creek, 21 – Lancefield and Warrnambool, 22 – Madigan Gulf, 23 – Mammoth Cave, 24 – Mootwingee, 25 – Penola, 26 – Port Augusta, Spencer Gulf and Dempsey's Lake, 27 – Salt Creek, 28 – Thorbindah, 29 – Warburton River and Cassidy Locality, 30 – Wellington Caves. Sites initially attributed to Dromornithidae: 31 – King's Creek, 32 – Scott River*

4.3.1. *Eocene*

4.3.1.1. *Redbank Plains*

The oldest fossil site having yielded dromornithids is Redbank Plains, situated close to Brisbane in Queensland (Figure 4.4), where a footprint cast was discovered and attributed to an individual of Dromornithidae *incertae sedis* (Murray & Vickers-Rich, 2004). This footprint was discovered in the Redbank Plain Formation and dates back to the Early Eocene (50–55 Ma), corresponding to the environment of a shallow lake where a significant fauna was discovered (Murray & Vickers-Rich, 2004).

4.3.1.2. *Old Endurance Tin Mine*

This is the only site containing remains of Dromornithidae known in Tasmania (in the north-east; Figure 4.4) (Spry & Banks, 1962). It corresponds to a group of several hundred footprints discovered in sediments dating back to the Early Eocene (Harris, 1968), attributed to these birds thanks to several studies (Rich & Green, 1974).

4.3.2. *Upper Oligocene (Lower Miocene)*

4.3.2.1. *Riversleigh Station*

This is the only Oligocene site, corresponding more precisely to the end of the Oligocene and the beginning of the Miocene (24–15 Ma), situated in a sedimentary unit called Carl Creek Limestone in north-western Queensland (Archer *et al.*, 1997; Travouillon *et al.*, 2006; Black *et al.*, 2012; Worthy *et al.*, 2014; Worthy & Yates, 2015) (Figure 4.4). This site has yielded the first fossils of *Barawertornis tedfordi* (Tedford, 1968), a second indeterminate dromornithid whose size was intermediate between that of *Ilbandornis woodburnei* and that of *Bullockornis planei* (nowadays *Dromornis planei*), and a large amount of bones from sites around Riversleigh (Archer *et al.*, 1994), very few of which have been described so far.

4.3.3. *Miocene*

4.3.3.1. *Snake Dam*

This fossil site has yielded a fragment of a particularly thick eggshell attributed to *Dromornis* sp. (Williams & Vickers-Rich, 1992). This site, located at the Muborina Station, south-east of Lake Eyre (Figure 4.4), represents a stratigraphic unit similar to and probably contemporaneous with the Etadunna Formation, which dates back to the Middle Miocene (Murray & Vickers-Rich, 2004).

4.3.3.2. *Bullock Creek*

Bullock Creek is located south of Katherine (in the northern part of the Northern Territory; Plane & Gatehouse, 1968) (Figure 4.4) and dates back to the Middle Miocene (15–10 Ma) (Murray & Megirian, 1992; Megirian *et al.*, 2010). It represents one of the most significant fossil sites and it has yielded several hundreds of specimens (Murray & Megirian, 1992; Megirian *et al.*, 2010; Worthy & Yates, 2015). The species *Bullockornis planei* (nowadays synonymized with *Dromornis planei*), named after this site and one of the discoverers, was found in abundance at this site, a large part of the specimens being very well preserved (Murray & Megirian, 1992).

4.3.3.3. *Alcoota Local Fauna*

Like Bullock Creek, this is one of the most significant sites containing Dromornithidae. It belongs to the Waite Formation (in the central part of the Northern Territory) (Figure 4.4) and dates back to the end of the Miocene (around 9–7 Ma) (Murray & Vickers-Rich, 2004; Megirian *et al.*, 2010). It represents the type locality for the genus *Ilbandornis*, but a large number of specimens of *Dromornis* have also been found there, including adult and young individuals with a considerable size range (Newsome & Rochow, 1964; Woodburne, 1967; Murray & Megirian, 1992; Murray & Vickers-Rich, 2004).

4.3.3.4. *Ongeva Local Fauna*

This site is Late Miocene in age and it is geographically close to the Alcoota Local Fauna (Figure 4.4), where the genera *Ilbandornis* and *Dromornis* have been identified (Woodburne, 1967; Murray *et al.*, 1993; Megirian *et al.*, 1996).

4.3.3.5. *Leaf Locality*

This fossil site, of Early Miocene age (23.4 Ma), is located east of Lake Ngapakaldi (Figure 4.4), where phalanges and a fragment of synsacrum associated with a diverse fauna have been discovered (Stirton *et al.*, 1967; Rich, 1979; Megirian *et al.*, 2010).

4.3.3.6. *Canadian Deep Lead Mine*

Canadian Deep Lead Mine is Middle Miocene in age (Dulhunty, 1971; Murray & Vickers-Rich, 2004) and is located close to Mudgee, in the mining district of Gulgong (Figure 4.4). A fragment of synsacrum was discovered there at the end of the 19th Century and sent to Owen to be studied (Owen, 1879a). Owen referred it to *Dromornis australis* but, given the state of preservation of the fossil, it is now considered as Dromornithidae *indet.* Later, a dromornithid femur was also discovered in this mine (Lydekker, 1891).

4.3.4. *Pliocene*

4.3.4.1. *Peak Downs and Lawson Quarry*

Only two fossil sites containing remains of Dromornithidae are known from the Pliocene of Australia. Among these, the Peak Downs site, in the eastern part of Queensland (Figure 4.4), has yielded the type specimen of *Dromornis australis*, which corresponds to a femur discovered about 55 meters underground in fluvial deposits, suggesting that it is Late Pliocene in age (Owen, 1874; Murray & Vickers-Rich, 2004). The second Pliocene site is Lawson Quarry, situated west of Lake Palankarinna (in the northern part of Southern Australia, Figure 4.4), where a dromornithid synsacrum was discovered (Miller, 1963; Murray & Vickers-Rich, 2004).

4.3.5. *Pleistocene*

The vast majority of sites containing remains of Dromornithidae date back to the Pleistocene and scattered around the whole southern half of Australia.

4.3.5.1. *Western Australia*

4.3.5.1.1. King's Creek, Scott River and Mammoth Cave

Three fossil sites are known in the Pleistocene of Western Australia. The King's Creek site (close to Darling Downs) (Figure 4.4) has only yielded a fragmentary femur that was first described as *Dinornis queenslandiae* (DeVis, 1884, 1891) and later reassigned to a moa, *Pachyornis elephantopus* (Scarlett, 1969), known from southern New Zealand, which must have been inventoried incorrectly (Rich, 1979). Scott River is a dune environment (Figure 4.4), where an egg, initially attributed to Dromornithidae, was discovered (Butler, 1969). However, this egg was at the center of many studies that unambiguously attributed it to Aepyornithiformes (see the chapter about Aepyornithiformes) rather than Dromornithidae (Long *et al.*, 1998; Long, 2002; Balanoff & Rowe, 2007). Thus, only the Mammoth Cave site, situated close to Witchcliffe (Figure 4.4) and dating back to the Pleistocene (37000 to 31500 years B.P.) (Lundelius, 1960; Tedford, 1967; Rich, 1979), yielded a single sacral vertebra that may be attributed to Dromornithidae and is part of a very abundant fauna of fossil vertebrates (Glauert, 1910, 1914, 1948).

4.3.5.2. *Queensland*

4.3.5.2.1. Diamantina River and Thorbindah

Diamantina River and Thorbindah (or Torlinda) are the only two Pleistocene sites known in Queensland (Figure 4.4) and dated on the basis of their fauna (Woods, 1960). A fragmentary tarsometatarsus and some phalanges, attributed to *Genyornis*, have been discovered at Diamantina River, even though the exact origin of these fossils, which may also come from Warburton River (Southern Australia) downstream from Diamantina River (Rich, 1979), is still uncertain. The Thorbindah fossil site lies along the Parro River in the southern part of Queensland, close to Cainwarra Station (Murray & Vickers-Rich, 2004), and has only yielded a tibiotarsus associated with a fibula and attributed to *Dromornis australis* (Etheridge, 1889). Unfortunately, the fibula is not a bird bone and the original femur has disappeared, so that only a cast remains. Despite the disappearance of the initial fossil and the poor preservation of the cast, this specimen was reassigned to *Genyornis newtoni* by Stirling and Zietz (1900). Given the material available, however, it is wiser to consider it as Dromornithidae *indet.* (Rich, 1979).

4.3.5.3. *New South Wales*

4.3.5.3.1. Wellington Caves, Cuddie Springs and Mootwingee

Wellington Caves (or Wellington Valley) is a group of limestone caves in the central part of New South Wales (Figure 4.4), where the first specimen (a femur) undoubtedly attributable to Dromornithidae was discovered (Mitchell, 1839), making this site especially significant from a historical point of view. It is difficult to date the remains of Dromornithidae because of a mixture of several sedimentary events in this cave. They were estimated to date back to the Pleistocene (Rich, 1979), but other studies are more cautious and suggest an age ranging from the Pliocene to the present (Tedford, 1966). In the absence of new data, it is more reasonable to suggest an age ranging from the Pliocene to the Pleistocene. The second site dating back to the Pleistocene (30000 years for some of the dated specimens, Field & Boles, 1998) is Cuddie Springs, situated close to Brewarrina (Gelgine Station) in the northern part of New South Wales (Figure 4.4). It contains fluvial deposits that have yielded several leg bone fragments of *Genyornis newtoni* (Anderson & Fletcher, 1934). Finally, the last site is Mootwingee, located north of Broken Hill in the western part of New South Wales (Figure 4.4), where a series of footprints that can be attributed to Dromornithidae was discovered at the same time as footprints possibly left by emus. This suggests the late presence of Dromornithidae and represents one of the arguments in favor of the cohabitation between Dromornithidae and Australian aboriginals (Murray & Vickers-Rich, 2004). Nonetheless, no detailed study or dating has been carried out, so that these footprints cannot be identified or dated with precision.

4.3.5.4. *Victoria*

4.3.5.4.1. Warrnambool and Lancefield

These are the only two sites known in the Victoria region (Figure 4.4). The Warrnambool site contains footprints that may be attributed to Dromornithidae, discovered in sand dunes that date back to the Pleistocene (Rich & Gill, 1976; Murray & Vickers-Rich, 2004). The Lancefield site, which also dates back to the Pleistocene (and more precisely to around 27000 B.C.) (Gillespie, 1978), has only yielded a fragmentary tarsometatarsus that can be attributed to *Genyornis newtoni* (Murray & Vickers-Rich, 2004). This site may be a former watering hole where a large number of animals, among which Dromornithidae, might have come to drink during severe droughts and died due to the lack of food and water (Horton, 1978).

4.3.5.5. *Southern Australia*

4.3.5.5.1. Callabonna Lake

This site dates back to the end of the Pleistocene (40000 years B.P. Stirling & Zietz, 1900; Tedford, 1971) (Figure 4.4) and represents the fossil site where the first and only subcomplete articulated skeleton of *Genyornis newtoni* was discovered in 1892 (Murray & Vickers-Rich, 2004) together with a large number of partially articulated and related dromornithid fossils, mainly attributable to the genus *Genyornis* and associated with an abundant fauna (Tedford, 1984; Stirling, 1896; Stirling & Zietz, 1900, 1913, 1896, 1905; Murray & Vickers-Rich, 2004).

4.3.5.5.2. Cooper Creek and its region

The Cooper Creek region (including Emu Camp (= Malkuni Waterhole) and Lower Cooper Creek Locality 2) and the site itself represent a group of several fossil sites dating back to the Pleistocene and in which fragmentary and scanty remains of Dromornithidae were discovered (Rich, 1979) (Figure 4.4).

4.3.5.5.3. Warburton River and Cassidy Locality

These Late Pleistocene sites have yielded some remains including the hindlimbs and a vertebra of a dromornithid referred to *Genyornis newtoni* (Rich, 1979). These sites lie along the Warburton River (Figure 4.4), but the fossil levels are unknown since the bone elements have been discovered on the surface.

4.3.5.5.4. Brothers Island

This site is located west-north-west of Port Lincoln (Figure 4.4). It has yielded a femur referred to *Genyornis newtoni* and it has been estimated to be Pleistocene in age thanks to the associated fauna (Rich, 1979).

4.3.5.5.5. Big Cave

This site lies in the farthest south-eastern part of Southern Australia (Figure 4.4) and is Pleistocene in age. The only element discovered in this site and attributable to a dromornithid corresponds to a fragmentary cervical vertebra that may be referred to *Genyornis newtoni* (Rich, 1979).

4.3.5.5.6. Gambier Range

This site is of more historical than scientific relevance. It was there that the lower part of a tibia was excavated and sent to Richard Owen to be studied. This is one of the specimens described by Owen (Owen, 1879) as belonging to the species *Dromornis australis*. However, it is most likely a tibiotarsus of *Genyornis newtoni*, as suggested later by Stirling and Zietz (1900). This site corresponds to a cave, probably Pleistocene in age, where fluvial deposits have accumulated (Lydekker, 1891) (Figure 4.4).

4.3.5.5.7. Penola

This Pleistocene site lies about 25 km north-north-west of the town of Penola on the border between Southern Australia and Victoria (Woods, 1866) (Figure 4.4). This site is interesting because Wood noted that the bones referred to the Dromornithidae discovered there might have been contemporaneous with the Aboriginals, since they revealed traces of traditional tools and fire. Wood suggested the name *Dromaius australis* for these specimens. However, he did not describe or illustrate them, so that the name was invalid. Unfortunately, Wood's descriptions are too succinct to reassign these fossils and the specimens themselves have now disappeared (Rich, 1979).

4.3.5.5.8. Salt Creek

Fossil material was discovered on the bed of a stream in Salt Creek, close to Normanville (Figure 4.4). These fossils, which corresponded to fragmentary leg bones, were attributed to *Genyornis* (mainly based on size) (Stirling & Zietz, 1900). Thanks to the study of the associated fauna, it was referred to the Pleistocene (Rich, 1979).

4.3.5.5.9. Baldina Creek

Baldina Creek is located close to Burra in the south of Southern Australia (Figure 4.4) and dates back to the Pleistocene (34100±1440 B.P.; Williams, 1981). Only a fragmentary femur was discovered there and briefly mentioned by Stirling and Zietz (1900) as referred to *Genyornis*.

4.3.5.5.10. The confluence of the Darling and Murray rivers

Eggshells were discovered at the confluence of the Darling and Murray rivers and attributed to *Genyornis newtoni* (Miller *et al.*, 1999) (Figure 4.4).

4.3.5.5.11. Port Augusta

Several hundred eggshell fragments of *Genyornis* were also discovered in Port Augusta (at Dempsey's Lake) in dune sediments dating back to the Pleistocene on the southern coast of Southern Australia (Williams, 1981) (Figure 4.4).

4.3.5.5.12. Hookina Creek

Hundreds of eggshell fragments were discovered in sediments dating back to the Pleistocene in Hookina Creek, which lies 100 km north-east of the region (Figure 4.4), and referred to *Genyornis* (Williams, 1981).

4.3.5.5.13. Spencer Gulf

In 2001, a subcomplete egg *in situ* was discovered by Nigel Spooner close to Spencer Gulf Point and also attributed to *Genyornis newtoni* (Grellet-Tinner *et al.*, 2016) (Figure 4.4).

4.3.2. *The attribution of non-osseous fossils to Dromornithidae*

4.3.2.1. *The attribution of Tertiary footprints to Dromornithidae*

Footprints have been identified at the sites of Mootwingee (Pleistocene, New South Wales), Redbank Plains (Lower Eocene, Queensland) and Old Endurance Tin Mine (Lower Eocene, Tasmania) (see above). Thus, these three sites are all Cenozoic in age, excluding the hypothesis that dinosaurs left these footprints. Moreover, no large three-digit animal has been found in Cenozoic sediments in Australia, whether mammal, reptile or amphibian, making the hypothesis that they were left by birds more likely (Rich, 1979). Among the large birds known from Australia, only ratites (emus and cassowaries) and Dromornithidae could have left these footprints. To determine which of these birds left them, a study was carried out on the footprints left by extant emus and cassowaries at the New York Zoo. Even though these two groups of present-day ratites share some features with the fossil footprints, cassowaries differ because of the elongate distal phalanx of their middle toe, which cannot be observed in the fossil footprints. On the other hand, the footprints left by emus are very similar to the fossil ones, but there are nevertheless a certain number of differences (Rich, 1979). Furthermore, if we compare the fossil footprints with those that may have been left by the *Dromaius* known to have lived in mid-Tertiary times, the former are clearly larger. Consequently, for the time being, it seems reasonable to attribute the fossil footprints found in Australia to Dromornithidae (Rich, 1979).

4.3.2.2. *The attribution of eggshells and eggs to Dromornithidae*

Apart from the entire eggs initially attributed to Dromornithidae, which turned out to be aepyornithid eggs, the very large number of eggshell fragments discovered in Australia were thought to derive from eggs laid by *Genyornis newtoni*. The first to refer these eggs to these birds was Williams (1981), who studied a large number of eggshells from the Pleistocene of the southern part of Southern Australia. His study led him to conclude that these fragments derived from an egg that must have weighed around 1.3 kg, which does not correspond to that of any extant Australian bird, and the fact that they were considered to be Pleistocene in age excluded the hypothesis that the eggs belonged to any of the birds imported by humans. Thus, the only birds that could have laid these eggs are Dromornithidae, and the only representative of this group still alive during the Pleistocene was *Genyornis newtoni*. Therefore, Williams (1981) suggested that all particularly thick eggshell fragments deriving from large eggs should be attributed to this species.

Later, Williams and Vickers-Rich (1992) carried out a second study on a single eggshell fragment coming from Snake Dam. Based on the very significant thickness of this eggshell (around 4 mm) and the study of its curvature, they concluded that it must have come from an especially large egg (with a size equal to or greater than that of *Aepyornis* eggs), whose weight must have ranged from 12 to 15 kg. As this fragment had a typically avian microstructure, despite being unique among Australian birds, they suggested that it should be referred to Dromornithidae and, more specifically, to *Dromornis* sp., which was the largest form of Dromornithidae not only from the Miocene but also in general.

However, the attribution of these eggshells to *Genyornis newtoni* has been recently challenged by the work carried out by Chan (2014) and Grellet-Tinner *et al.* (2016), who have shown using different methods the great difference in size between the estimated size of the eggs (through the study of the shells or the subcomplete egg) and the size of the bones of *Genyornis*. Chan (2014) has shown that the mass that the eggs could withstand during incubation was much smaller than the mass of adult *Genyornis*, thus challenging the initial hypothesis that attributed these eggs to these birds. According to these conclusions, adult *Genyornis* would have crushed and squashed their eggs during incubation if these were actually the eggs from which the Pleistocene shells originated. Subsequently, these conclusions were confirmed by the work carried out by Grellet-Tinner *et al.* (2016), who studied the microstructure of a large number of eggshells, associated with chemical analyses of the shells as well as estimates of the mass of the eggs based on the mass of *Genyornis* and vice versa, showing thus that *Genyornis* could not have laid these eggs, which are too small given the size of these birds. Therefore, they have suggested that these eggs might have been laid by Australian megapodes.

Given the available evidence, the identification of the shell found at Snake Dam as *Dromornis* is not challenged, but it seems evident that a single shell is not enough to put forward a definite and reliable conclusion. The case of the other shells is still unclear as well. Even if the studies carried out by Chan (2014) and Grellet-Tinner *et al.* (2016) have shown that part of these shells (all coming from the same region and corresponding to those studied by Williams (1981) and those found in the Port Augusta (Chan, 2014), Spencer Gulf and Dempsey Lake sites (Grellet-Tinner *et al.*, 2016)) cannot be attributed to Dromornithidae, all the shells found at the other sites (among which those found in the Lake Eyre basin) can so far still be attributed to *Genyornis newtoni* (Grellet-Tinner *et al.*, 2016).

4.4. Anatomy

Figure 4.5. *Skeleton and restoration of* Dromornis stirtoni

The size and body mass of Dromornithidae vary widely, ranging from 1.5 m and 60 kg to 3 m and 550 kg, and their anatomy has been much studied, as summarized by Murray and Vickers-Rich (2004) (Figure 4.5). Additional studies are still ongoing (Nguyen *et al.*, 2010; Worthy *et al.*, 2014; Worthy & Yates, 2015).

The reduction of flight-specific attributes is very significant in these birds, with a sternum that forms a non-carinated bone plate, very small wings, and scapula and coracoid fused into a non-mobile unit.

The beak of the largest Dromornithidae (*Dromornis planei* and *Dromornis stirtoni*) is high and presents significant dorsal ridges. Despite its solid appearance, it is actually lightly built and made mostly of spongy bone (Worthy *et al.*, 2016). On the contrary, the smaller forms (such as *Genyornis newtoni*) have a much smaller beak which is much more similar to that of extant geese. The eyes of the largest individuals are very small and their mandible is very large in comparison with the size of their skull (Figure 4.6).

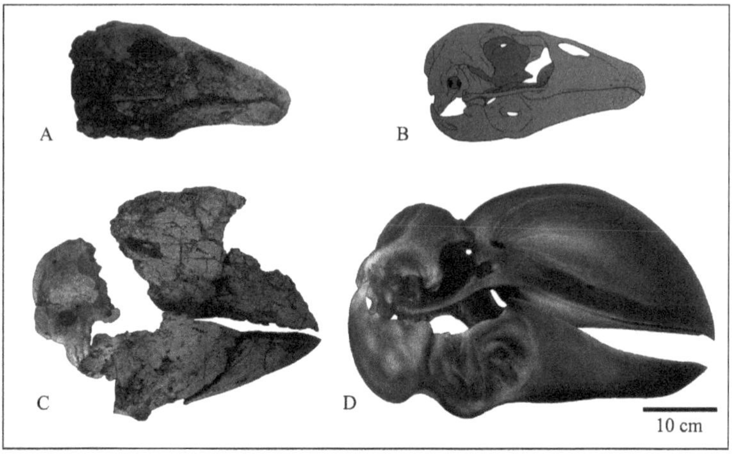

Figure 4.6. *Examples of dromornithid skulls. (A) Skull of* Genyornis newtoni *(from Stirling & Zietz, 1913), (B) Restoration of the skull of* Genyornis newtoni *(from Murray & Vickers-Rich, 2004), (C) Skull of* Dromornis planei *(from Murray & Vickers-Rich, 2004), (D) Restoration of the skull of* Dromornis planei *(from Murray & Vickers-Rich, 2004)*

The leg bones of the largest types of Dromornithidae are very large in relation to those of present-day "geese" due to their very large size, but the proportions of the bones are quite similar to those seen in some extant "geese" such as *Anseranas semipalmata*. On the other hand, the species *Ilbandornis lawsoni* has a long and thin tarsometatarsus which is very different from that of other Dromornithidae (Murray & Vickers-Rich, 2004; Worthy & Yates, 2015) (Figure 4.7). Thanks to the footprints discovered especially in the Old Endurance site in Tasmania (see above), it has been possible to study the position of the digits while the animal was moving. Thus, it can be noted that the hind toe (or digit I) is missing. All three digits are relatively short, nearly uniformly and widely separated, with a central digit that is longer than the others and the two lateral digits of a similar size with a maximum difference of 4 cm in length. The outline of the footprints is generally round, with a rounded and wide terminal phalanx, more similar to hoofs than claws.

While the anatomy of the hindlimbs is mainly dictated by biomechanical constraints associated with size, some features are present independently of the size of the species, such as particularly short and wide toes blunted at the end with a terminal phalanx similar to a hoof (Figure 4.7). We can also mention the presence of grooves visible on the margin of the upper jaw or a complex structure joining the mandible to the skull. The rest of the anatomy has proportions fairly similar to those of present-day geese with a long neck, a wide pelvis and a long tibiotarsus.

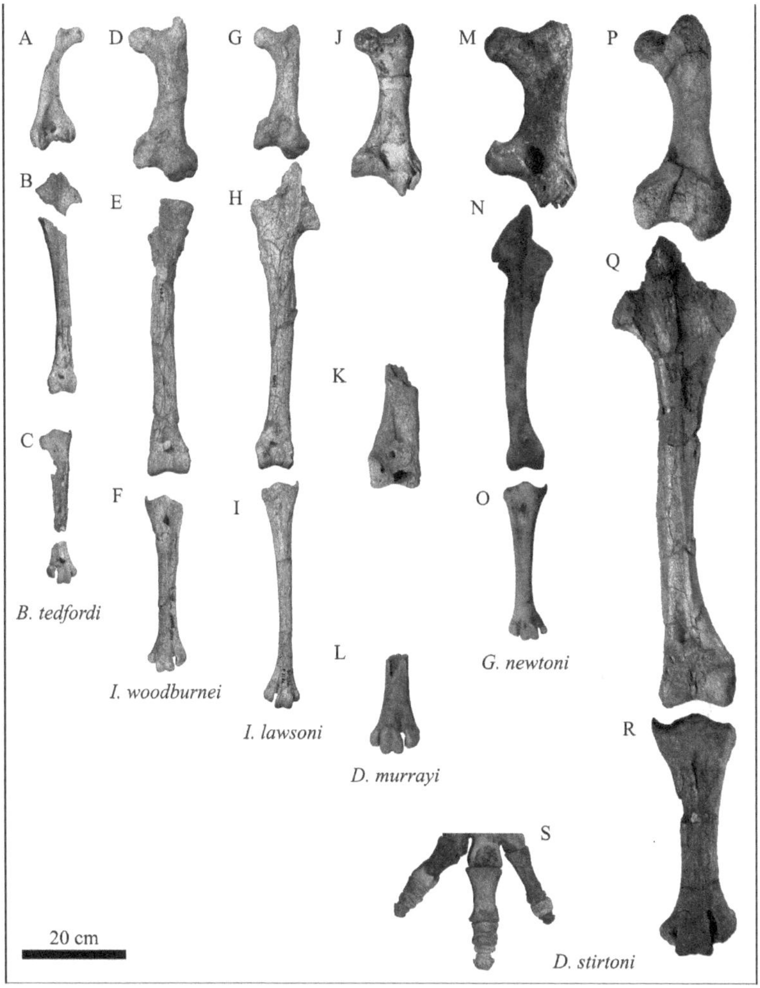

Figure 4.7. *Bones of the hindlimbs of some Dromornithidae. A, D, G, J, M, N: femurs; B, E, H, K, N, Q: tibiotarsi; C, F, I, L, O, R: tarsometatarsi; Q: phalanges. Images A-C from Nguyen* et al. *(2010), images D-I from Worthy & Yates (2015), images J-L from Worthy* et al. *(2016) image M from Grellet-Tinner* et al. *(2016), images P-R from Handley* et al. *(2016) and images, N, O, Q drawn by Murray & Vickers-Rich (2004);* B. tedfordi*:* Barawertornis tedfordi, I. woodburneyi*:* Ilbandornis woodburneyi, I. lawsoni*:* Ilbandornis lawsoni, D. murrayi*:* Dromornis murrayi, G. newtoni*:* Genyornis newtoni, D. stirtoni*:* Dromornis stirtoni. *For a color version of the figure, see www.iste.co.uk/angst/birds.zip*

4.5. Paleophysiology and paleoecology

4.5.1. *Diet*

As is the case for Gastornithidae, the very peculiar shape of the mandible of Dromornithidae makes it very hard to assess their diet by simply comparing it with that of present-day birds. This is all the more true as mandibular remains have not been found for all species of Dromornithidae and the existing fossils are very poorly preserved. For example, the beak of *Ilbandornis lawsoni* or the mandible of *Dromornis australis* are currently completely unknown, whereas the skull and the mandible of *Genyornis newtoni* are known but the bones are so fractured that a precise anatomical study cannot be carried out (Murray & Vickers-Rich, 2004; Worthy *et al.*, 2016). The diet of Dromornithidae, therefore, is much debated.

Wroe (1999) suggested that these birds were carnivores as their jaws and their cranial muscles were "oversized" for a herbivorous bird, as Witner and Rose (1991) had previously hypothesized for Gastornithidae. Wroe thought that these birds, given their large and powerful jaws, could not merely feed on plants. The main argument of these authors is that an increase in head size in order to feed merely on plants, which provide a limited energy intake, presents no evolutionary advantage. Thus, when a bird grows larger in the course of evolution, the size of its head need not grow proportionally, as can be seen, for example, in extant ostriches, which are herbivorous and have a proportionally small head in relation to their size. Wroe (1999), in line with Witmer and Rose's arguments (1991), suggested then that Dromornithidae must be carnivores, with two possibilities: either these birds hunted their prey and dismembered it, or they were scavengers and used their incredibly powerful jaws to crush bones.

However, this hypothesis and all the arguments put forward by Wroe were refuted by Murray and Vickers-Rich (2004). Thus, the evolution of large herbivorous birds in the direction of a proportionally smaller head is not supported by any measurement or objective element and cannot be proven for all herbivorous birds (Murray & Vickers-Rich, 2004). Moreover, Murray and Vickers-Rich (2004) showed that ostriches, and ratites in general, have an anatomically unique skull, which limits their abilities to handle food and consequently the range of food to which these birds have access, which was not the case for Dromornithidae, whose cranial muscular system was highly developed. Thus, if Dromornithidae were herbivores, they must very likely have had access to a wide range of plants with a consequently more substantial nutritional value. Another argument against a carnivorous diet is the absence of claws and a hooked beak in Dromornithidae, which are, however, used by all extant carnivorous birds not only to immobilize and kill their prey, but also to rip flesh apart once the animal is dead (Murray & Vickers-Rich, 2004). Furthermore, Murray and Vickers-Rich (2004) add that the eyes of large Dromornithidae such as *Dromornis* are very small, situated laterally on the skull, and associated with very small optic lobes, which is not the case for most present-day birds of prey. Studying the shape and size of the jaw of these birds also shows that it would not allow them to crush or swallow large pieces of bones they may have crushed. Similarly, their very large and massive beak could not have allowed these birds to feed on the flesh left on carcasses between the ribs or the vertebrae, making scavenging difficult, to say the least.

Therefore, the hypothesis currently accepted is that Dromornithidae were herbivorous birds. Large forms such as *Dromornis* must have fed on a wide range of plants, more or less tough, thanks to their strong and highly mobile jaw (which they did not use to crush the bones of their prey, but to squash seeds and fruits) (Murray & Vickers-Rich, 2004). The fore part of their beak must have allowed them to grab leaves, flowers, bulbs and fruits by retracting their neck, their long beak was useful for choosing fruits, and their high and strong jaws must have allowed them to grab branches. The intermediate part of their jaw must have been used to crush food. To meet their daily energy needs, Dromornithidae likely consumed a large share of fruits, nuts and seeds, which are highly caloric (Murray & Vickers-Rich, 2004). Thus, it has been suggested that these birds played a significant role in the dispersal of a large number of plants, on which they fed.

The case of *Genyornis* is slightly different. Despite the very poor preservation of the skull of this bird, it can be observed that the beak and jaws of this dromornithid were not as strong as those of *Dromornis*, and that it was more similar to present-day geese. By analogy, a grazing herbivorous diet is then the most convincing hypothesis (Murray & Vickers-Rich, 2004). This was verified by Miller *et al.* (1999) through the carbon isotope analysis of the eggshells attributed to *Genyornis*, which allowed them to demonstrate that these birds mainly fed on C3 plants.

Another element worth mentioning is the presence of gastroliths, which have been described in the stomach of several specimens of Dromornithidae, among which *Genyornis*, *Dromornis* and *Ilbandornis* (Stirling & Zietz, 1905; Tedford, 1984; Williams, 1981; Murray & Vickers-Rich, 2004). It is still being debated (as is often the case) whether these gastroliths are real or small gravel naturally eroded and polished without having passed through a bird's stomach. Nonetheless, it should be noted that the majority of these stones consist of quartzite grains, whereas the surrounding sediments were derived from metamorphic and crystalline rocks, which seems to confirm the hypothesis in favor of actual gastroliths. These quartzite stones are smooth, round and often shiny, with a diameter ranging from 1 cm to 5 cm (Murray & Vickers-Rich, 2004).

4.5.2. *Body mass*

The body mass of Dromornithidae was estimated by Murray and Vickers-Rich (2004) and then by Nguyen *et al.* (2010) with different methods (Table 4.1). However, it should be noted that the Campbell and Marcus equation (1992) used by Murray and Vickers-Rich (2004) has been incorrectly transformed mathematically and is therefore false, invalidating part of their body mass estimates. On the other hand, as these authors have provided part of the initial measurements of the minimum circumferences of the femora they have measured, we have been able to re-estimate body mass values based on the right Campbell and Marcus equation (1992) (Table 4.1). It should also be noted that part of the average body mass values estimated by Nguyen *et al.* (2010) are also false, even though their minimum and maximum values are correct. *Barawertornis tedfordi* corresponds to the lightest form of Dromornithidae weighing around 60 kg, whereas *Dromornis stirtoni* is clearly the heaviest, with an average body mass of around 550 kg. *Dromornis stirtoni* also slightly differs from other species, as it represents the only species that presents a strong sexual dimorphism (see below), resulting in males that are 80 to 140 kg heavier than females (Handley *et al.*, 2016). The femur of this very heavy dromornithid tends to have a wider diaphysis for biomechanical reasons, which leads to overestimation of its weight when using the Campbell and Marcus method (1992) based on the minimum circumference of the femur. This is why Handley *et al.* (2016) suggest that values estimated on the basis of the tibiotarsus should be preferred.

Methods	Campbell & Marcus (1992) Femur (kg)	Campbell & Marcus (1992) Tibiotarsus (kg)	Anderson *et al.* (1985) Femur (kg)	Alexander (1983) (kg)	Hurlburt (1999) (GDI model) (kg)	Others (kg)
Genyornis newtoni	409 (374-444) [1*] (327-444) [3]	(168-253) [3]	429 (387-470) [1] 259 (150-445) [2]		292 [1]	
Ilbandornis lawsoni	134 (116-164) [1*-2]	106 [2]	121 (103-153) [1]			
Ilbandornis woodburnei	211 (177-279) [1*] 225** (177-279) [2]	114 (101-131) [2]	203 (166.5-278) [1]			
Ilbandornis sp.	129 (126-131) [1*]		116 (113-118) [1]			
Dromornis stirtoni	545 (342-714) [1*] 519** (342-714) [2] 501.5 (303.9-744.5) [4]	491 (450.5-568) [2] 477 (348.9-632.1) [4]	587 (350-805) [1] 305.5 (190.5-444.3) [4]	786 [1]	570 [1]	491(298-728) [4] ♂ : 584 (439-728) [4] ♀ : 441 (317-560) [4]
Dromornis planei	333 (255-382) [1*] 331** (255-382.5) [2]		290 (251-397) [1]			
Dromornis australis	208 [1*]		200 [1] 133 (80-223) [2]			
Barawertornis tedfordi	(50-71) [1] 59 [2]	64 [2]	46 (29-73.5) [2]			

Table 4.1. *Compilation of the body masses in kilograms estimated for Dromornithidae in the literature. Mean value (minimum value – maximum value). *: values recalculated based on the raw data from Murray & Vickers-Rich (2004), ** false mean value; ♂: males, ♀: females; [Bibliographic references]: 1 – Murray & Vickers-Rich (2004), 2 – Nguyen* et al. *(2010), 3 – Chan (2014), 4 – Handley* et al. *(2016).*

Studying the evolution of the body mass of the largest Dromornithidae over time shows that the body mass of these birds increased substantially between the end of the Oligocene and the end of the Miocene, reaching a weight of around 500 kg for *Dromornis stirtoni*. Murray and Vickers-Rich (2004) have interpreted this increase in body mass as an ecological adaptation to environmental changes. Large Dromornithidae were adapted to a diet that included mainly fruits, which are more caloric and allowed these large birds to meet their daily needs. Seasonality progressively increased in Australia, making fruits also more seasonal and forcing large Dromornithidae to feed more on leaves or grass, which are less caloric. To counterbalance the energy deficit due to this change in their diet, these birds increased their size and their body mass (as the increase in size can reduce the surface/volume ratio and thus limit the energy spent regulating body temperature by reducing the relative surface of the body exposed directly to external temperature). However, at the end of the Miocene, unlike contemporary marsupials whose body mass had also increased, Dromornithidae reached a plateau and then saw their body mass decrease (Figure 4.8., Murray & Vickers-Rich, 2004). These authors associate this size constraint in Dromornithidae with the maximum size limit of the eggs they laid, as explained by several studies (Rahn *et al.*, 1975, 1985). The larger the eggs, the thicker the shell, until the shell reaches a threshold value past which a chick can no longer break it when hatching (Calder, 1984; Schmidt-Nielsen, 1985; Alexander, 1989). From the Early Pliocene onwards, while the body mass of marsupials kept increasing, that of Dromornithidae decreased. This may be explained by the fact that marsupials were able to keep increasing their size and body mass in response to the weak caloric nature of their diet, whereas birds were not able to do so due to the limiting size of their eggs (Murray & Vickers-Rich, 2004). The significant competition between large marsupials and large birds was then detrimental to Dromornithidae, only the smallest forms of which could subsist.

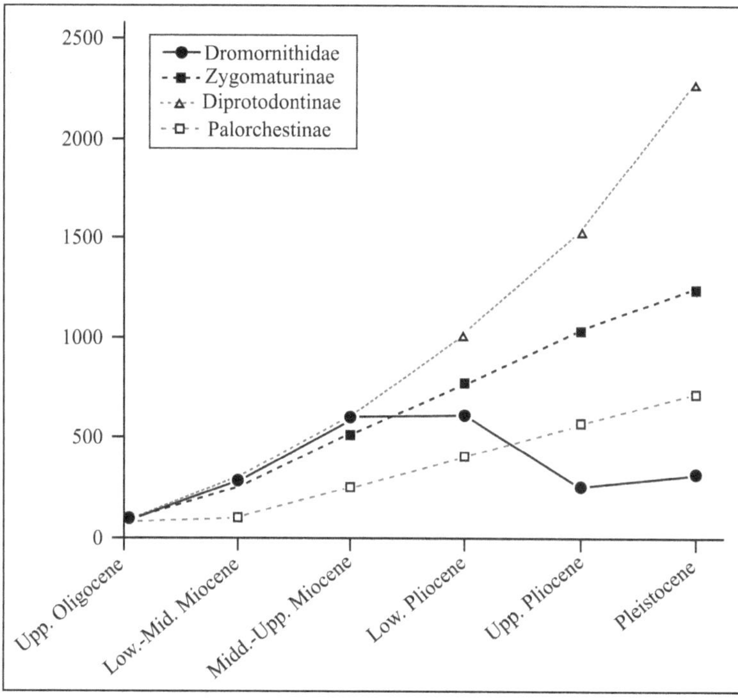

Figure 4.8. *Evolution of the body mass of Dromornithidae and various large marsupials over time (from Murray & Vickers-Rich, 2004)*

4.5.3. *Locomotion*

The first to deal with the locomotion of Dromornithidae were Stirling and Zietz (1900), who studied the anatomy of the phalanges of these birds and attributed their short shape and the presence of hoof-shaped terminal phalanges to a slow type of locomotion, by analogy with mammals. Claws can make it easier to have purchase on the ground while running and make up for lateral movements, which "hoofs" cannot do (Murray & Vickers-Rich, 2004). Locomotion was also investigated with the classic anatomical study of hindlimb bones (as it was clear that Dromornithidae, given their significantly reduced wings, could not fly). Thus, the very large femur of *Genyornis* must have allowed it to have a significant acceleration and deceleration force while walking quickly or running (Murray & Vickers-Rich, 2004). On the other hand, the bones of the hindlimbs of large Dromornithidae such as *Dromornis stirtoni* were very robust, suggesting a slower form of locomotion for these birds, whereas *Dromornis planei* had thinner and slenderer leg bones, which may be interpreted as a slightly more pronounced adaptation to running (Murray & Vickers-Rich, 2004). An extreme adaptation to running can be seen in *Ilbandornis lawsoni*, whose leg proportions are remarkably similar to those of present-day ostriches and emus, with long and thin bones, suggesting a completely cursorial type of locomotion for these birds (Murray & Vickers-Rich, 2004).

These interpretations based on the anatomical study of Dromornithidae were confirmed by a study of the length/width ratio of the tarsometatarsus (Angst *et al.*, 2016). The vast majority of Dromornithidae had a clearly graviportal type of locomotion corresponding to length/width ratios for their tarsometatarsus that range between 6.2 and 10.3, with a trend towards less graviportal locomotion for smaller forms such as *Genyornis newtoni* or *Ilbandornis woodburnei*. However, there is an exception among Dromornithidae, viz. the species *Ilbandornis lawsoni*, which has a tarsometatarsus anatomically very similar to that of present-day ostriches, with a ratio of 17.7, suggesting a clearly cursorial type of locomotion (Figure 4.9).

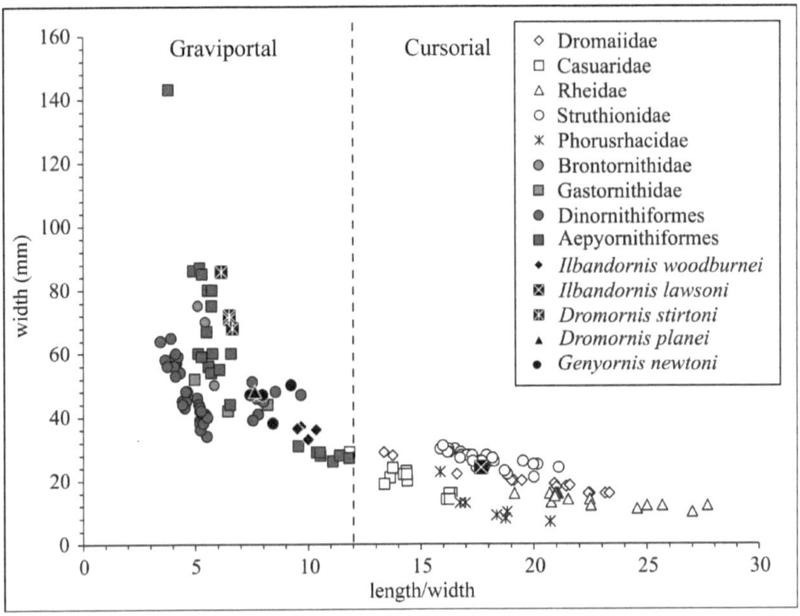

Figure 4.9. *Estimate of the locomotion type of Dromornithidae based on the method of Angst* et al. *(2016)*

Studying the locomotion of these birds was also greatly facilitated by the analysis of the footprint tracks attributed to these birds and discovered in the northern part of Tasmania (see above), which constitute direct evidence of their locomotion (Rich & Green, 1974; Rich, 1979). By measuring the size of the stride, it is possible to estimate the speed (Alexander, 1989; see Introduction). Thus, taking into consideration that the distance between two footprints ranges between 52.0 cm and 62.5 cm (Rich, 1979), the distance between two footprints left by the same foot may be estimated at around 1.15 m. By comparing these footprints and the bones of the foot of *Ilbandornis woodburnei*, Murray and Vickers-Rich (2004) suggested that these footprints should be attributed to this bird. Therefore, the length of the hindlimb (at hip height in a standing and resting position) is estimated to be around 88 cm, so that we can estimate the relative length of the stride to be around 1.3 m and thus obtain a walking speed of 5.2 km/h for this Dromornithidae (which corresponds to a normal walking speed for a human being).

There is no track left by Dromornithidae which indicates rapid running. However, Murray and Vickers-Rich (2004) have modeled the maximum range of motion allowed by these birds' hindlimbs based on leg bones. Thus, according to them, *Dromornis stirtoni* could easily walk at 10 km/h and run at around 17 km/h when stretching their stride to its anatomical limit. By adopting the same approach, these authors have also been able to estimate that *Ilbandornis lawsoni*, the dromornithid whose legs are most anatomically similar to those of ostriches, must have been able to run as fast as 55 km/h.

Like body mass, the locomotion type of Dromornithidae has evolved over time, from a rather graviportal to an increasingly more cursorial type of locomotion. Murray and Vickers-Rich (2004) suggested that the ability of some Dromornithidae to run fast was linked to a nomadic lifestyle or migration phenomena and to a more open and drier environment. This may also be associated with the aridification that took place in the Middle Miocene, which potentially reduced the amount of food or water locally available and made it necessary to move over significant distances to access new supplies.

4.5.4. *Salt gland*

On the skulls of several Dromornithidae (*Dromornis*, *Ilbandornis* and *Genyornis*), a long depression can be seen on the dorsal edge of the orbit that communicates with the inner orbit (Murray & Vickers-Rich, 2004). This type of structure corresponds to a salt gland, an organ which has been identified in several present-day marine birds, and also in non-marine birds living in desert environments (Dorst, 1982). These glands can remove excess salt from the birds' blood, thus allowing them to drink only saltwater or brackish water (Schmidt-Nielsen, 1985). In Dromornithidae, the presence of channels that can excrete the salty liquid filtered by salt glands demonstrates how these glands played an active role in these birds and did not merely constitute a potential form of phylogenetic heritage.

These functional salt glands, therefore, must have helped Dromornithidae adapt to the arid conditions of the Miocene and Pleistocene. As the size of the salt glands was more or less proportional to the amount of saltwater drunk by the birds, the significance of this structure for Dromornithidae challenges the idea that these birds must have lived in a tropical forest environment, where this structure would not have been useful for them and would have consequently regressed or disappeared. Dromornithids, therefore, must have been able not only to drink more or less salty water and thus to resist dehydration, but also to consume plants with a higher degree of salinity, which allowed them to access a wider range of food.

4.5.5. *Reproduction*

4.5.5.1. *Eggs and juveniles of Dromornithidae*

A large number of eggshell fragments were discovered in Australia in Miocene and Pleistocene sites. Part of them were reattributed to megapodes (see above) (Grellet-Tinner *et al.*, 2016). On the other hand, the fragment discovered at Snake Dam and described in detail by Williams and Vickers-Rich (1992) is still attributed to *Dromornis*, while the shells found in the dune deposits of the Lake Eyre basin are still attributed to *Genyornis newtoni* (Miller *et al.*, 1999). However, only the *Dromornis* eggshell fragment was studied in detail. The Miocene shell from Snake Dam is 3.8 to 4.1 mm thick and, thanks to the equation devised by Ar *et al.* (1974),

Williams and Vickers-Rich (1992) were able to estimate that it should correspond to an egg ranging from 12.6 to 14.7 kg and from 24 to 29 cm in width and 30-34 cm in length, which makes it one of the largest eggs currently known.

The study of the rare bones of juvenile dromornithids allowed Murray and Vickers-Rich (2004) to establish the precocial nature of these fledglings, which must therefore have been autonomous and able to run away from danger very quickly after their birth, as is the case for present-day ostriches. However, Handley *et al.* (2016) suggested a K-reproduction strategy for these birds, implying that young individuals must have been cared for after by adults (see below).

Murray and Vickers-Rich (2004) also suggested that the growth of these chicks must have been very rapid after hatching, allowing them to reach a near-adult size in a few months, as is the case for present-day ostriches. Nonetheless, due to the lack of histological studies, this hypothesis is difficult to confirm for now. These authors also hypothesize that the fledglings must have had a more varied and omnivorous diet than adult individuals, including animal sources to sustain this very rapid growth. Once again, however, no actual analysis has confirmed this hypothesis, and supplementary studies will be required to test it.

4.5.5.2. *Sexual dimorphism*

The study of the sexual dimorphism of Dromornithidae is relatively incomplete, mainly due to the limited number of specimens known for most species, so that it is impossible to carry out a statistical study required to separate the sexes. Thus, the only two species studied were *Genyornis newtoni* and *Dromornis stirtoni*. Chan (2014) showed that *Genyornis* presented no sexual dimorphism. On the other hand, the species *Dromornis stirtoni* presents a very high degree of sexual dimorphism, which was mentioned very briefly for the first time by Murray and Vickers-Rich (2004). More recently, Handley *et al.* (2016) confirmed this observation by using different methods such as direct measurements of the bones, clustering, body mass estimates, and the use of landmarks and geometric morphometry. Furthermore, thanks to bone histology, they could demonstrate that the smallest individuals were females because of the presence of medullary bone. Sexual dimorphism may be due to the need of sharing ecological niches in an environment where food supplies are limited or it may depend on sexual selection. The population of *Dromornis stirtoni* studied by Handley *et al.* (2016) comes from the Alcoota site, where food supplies must have been enough in the Miocene to feed all the population. This is why these authors tend to interpret this significant sexual dimorphism as associated with a phenomenon of sexual selection.

4.5.5.3. *Sex ratio*

As they could distinguish between male and female individuals of *Dromornis stirtoni*, Handley *et al.* (2016) suggested that the sex ratio of the population of these birds should be studied. Thus, they compared the proportion of tibiotarsi attributed to females (based on the presence of medullary bone and differences in size and bone structure) with those attributed to male individuals. These bones were chosen based on several factors: the robustness and resistance of the bones during fossilization, and the large number of tibiotarsi of *Dromornis stirtoni* known, better reflecting the actual sex ratio of the population. The male:female ratio is then 1.1:1, suggesting that the number of males was slightly higher than the number of females, as is the case for some primitive Australian Anseriformes and some ratites (Worthy, 2009). Handley *et al.* (2016) interpreted this ratio as an element in favor of the hypothesis that *Dromornis stirtoni* must have had a mating system similar to that of Anserinae and characterized by long-term monogamy and the fact that both parents take care of their chicks and defend the nest they share.

4.5.5.4. *Incubation*

By identifying the sex of individuals of *Dromornis stirtoni*, Handley *et al.* (2016) have also been able to study the incubation behavior of these birds. To avoid crushing the eggs, only the lightest parent broods when there is significant sexual dimorphism. Therefore, only the female individuals of *Dromornis stirtoni* must have sat on the eggs, as they were significantly smaller and lighter than the males, unlike the behavior of moa (Worthy & Holdaway, 2002) or extant ratites (Davies, 2002).

Furthermore, based on the weight of the egg attributed to *Dromornis*, the incubation period could then be estimated thanks to the equation devised by Rahn and Ar (1974) and was found to last between 95 and 99 days. As this time span is especially long, Murray & Vickers-Rich (2004) put forward the hypothesis that these eggs must have been brooded in common with adults belonging to the community taking turns so that females could feed in turn.

They also hypothesized that nests must have included a maximum of eight or nine eggs, which must have been the maximum size that adult female bodies were able to cover.

4.5.5.5. *Reproduction strategy*

Very few juveniles or young adults have been found in the Alcoota site where, however, ovulating females have been described (Handley *et al.*, 2016). This observation may be explained in different ways:

– either it reflects a taxonomic bias, as the bones of juveniles or young adults are more fragile and have fossilized less well;

– or it is due to the fact that juveniles, growing extremely rapidly, reached their adult size in a few months or at most in two years, so that juvenile or young individuals were rarer;

– or young individuals did not grow up at this site, in line with a different geographical distribution according to the life history of *Dromornis stirtoni*, as had been suggested for *Dinornis robustus* (Holdaway & Worthy, 1997);

– or young individuals were simply rarer among the population, as females did not lay many eggs at once.

The last hypothesis may be explained by a K-reproduction strategy, which is observed in populations consisting of large-sized individuals with a long lifespan, late sexual maturity and limited fertility, leading to a population with few young individuals which are, however, taken care of by their parents, as has been suggested for moa (Handley *et al.*, 2016). This type of reproduction is favored by a predictable life environment, especially in terms of water and food supply, and limited risks. This reproduction strategy for *D. stirtoni* could potentially explain the reduction in size of Dromornithidae during the Miocene, which may be linked to the aridification of Australia in this period, making environmental conditions less favorable. Thus, the K-reproduction strategy of large Dromornithidae could have become a disadvantage in comparison with smaller Dromornithidae whose reproduction strategy may have been slightly different.

4.6. Paleoenvironment

Dromornithidae lived for more than 25 Ma in Australia and withstood and adapted to different types of environments, including different types of fauna and flora. The Australian environments of the Middle Eocene (50 Ma) were very humid, closed and relatively homogeneous, and included subtropical forests divided into hot tropical areas and cooler and more temperate areas. At the end of the Eocene (40 Ma), the leaching of soils, the advent of a colder climate and the strong aridification of the environment caused an increase in forest fires (MacPhail, 1995), which became more frequent during the Tertiary, thus opening the environment to wide grasslands of annually growing herbaceous plants and increasing the intensity of the fires. Therefore, a chain reaction led to the development of an important imbalance in the flora, with the creation of increasingly larger surfaces occupied by open environments and the selection of tree species resistant to or tolerant of fire (scleromorphic plants) such as eucalypts and acacias (indicated by pollen analysis), until they became predominant at the end of the Pleistocene (MacPhail, 1997; Murray & Vickers-Rich, 2004; Martin, 2006; Handley *et al.*, 2016).

Dromornithidae are known to have lived in widely different environments, both arid in the center of the continent and more humid in the south, including marshy environments and humid forests in Tasmania. These large birds coexisted with a diverse fauna that included mostly marsupials (such as kangaroos, Tasmanian devils or animals superficially similar to lions, antelopes, tapirs or rhinoceroses) or herbivorous land birds (such as emus), whereas carnivores were mainly represented by reptiles such as crocodiles, snakes or large monitor lizards (similar to Komodo dragons). Herbivorous marsupials were initially smaller than Dromornithidae, and none of them present any adaptation involving a long neck (like giraffes) that could enable a vertical distribution of ecological niches. During the Oligocene, these herbivorous marsupials evolved and their size and body mass became larger in order to adapt to the new Australian plants that provided less energy, so that some Diprotodons weighed more than 700 kg at the beginning of the Pliocene, clearly surpassing the largest Dromornithidae.

4.7. Extinction

Dromornithidae started declining in both size and diversity at the end of the Miocene (Murray & Vickers-Rich, 2004). The last species of Dromornithidae known is *Genyornis newtoni*, which lived until the Pleistocene. The study carried out by Miller *et al.* (1999) on eggshells of *Genyornis* demonstrated that these birds became suddenly extinct 50000 years ago as the first human beings arrived on the northern coast of Australia. Nonetheless, the extinction of this family is probably due to the combination of two factors: climate changes that severely altered the environment and then the arrival of the first humans in Australia, who probably delivered the final blow to these already declining birds.

4.7.1. *Climate changes*

Climate variations have certainly played a significant role in the decline of Dromornithidae. From the Middle Miocene onwards, a phenomenon of aridification can be observed both on a geomorphological and on a botanical level (Murray & Vickers-Rich, 2004). This altered the plant cover, especially with the increase in fires (due to increased dryness and thunderstorms), which favored the spread of grass and scleromorphic plants (quite resistant to fire), to the detriment of trees that were more sensitive to fire and represented the indigenous Australian flora. Moreover, the aridification of the environment made rainfall less predictable and, consequently, the production of fruits became less certain and more limited. This reduced the availability of energy-rich fruits produced by plants sensitive to fire, which must have represented a substantial part of the diet of at least some Dromornithidae, to which they were well adapted. This also increased the distance between the zones of indigenous forest.

To face these environmental changes, Dromornithidae adopted two strategies from the end of the Miocene onwards: they increased their size and body mass to make the most of the low-energy food supply of their new environment (e.g. *Dromornis stirtoni*), or they increased their locomotion speed to cover more distance and access a larger area where they could find their food (e.g. *Ilbandornis lawsoni*). However, these two strategies did not work for several reasons. Initially, the strategy based on the increase in body mass allowed Dromornithidae to occupy a different ecological niche from that of kangaroos and emus until the beginning of the Pliocene, as the former were significantly bigger than the latter. The increase in body mass reached its maximum with *Dromornis stirtoni*, as these birds could not increase their size and body mass anymore for metabolic reasons and due to the limiting size of the eggs (see above). From the end of the Pliocene and the beginning of the Pleistocene onwards, the size of Dromornithidae decreased (up to *Genyornis*), whereas kangaroos Diprotodons and emus became at the same time increasingly larger and often found themselves in direct competition to access the same type of food (initially shared according to size and therefore the height at which these animals could reach their food). Moreover, at equal body mass, these animals were in a better metabolic position than birds to use the energy derived from their food, allowing them to make do with less energy-rich fruits or leaves produced by the new flora and to avoid any constraints on their increase in body mass. Finally, the decreasing amount of fruits caused by this change in the flora could also lead Dromornithidae to adopt a diet that included more leaves, putting them in direct competition with other groups of leaf-eating herbivorous animals, which were better adapted to this kind of diet. As for the adaptation strategy of Dromornithidae towards increased speed, the geographical distribution of the indigenous flora became too discontinuous and the production of fruits too irregular and unpredictable to allow these birds to survive.

4.7.2. *Human action*

The extinction date of Dromornithidae shows that they coexisted for several thousand years with the first humans who arrived in Australia between 100000 and 60000 B.P. years (Murray & Vickers-Rich, 2004). Nevertheless, nowadays it is still difficult to determine whether these birds interacted with human beings and, if this is the case, how. Few elements allow us to confirm that there have actually been direct contacts between the last Dromornithidae and the first aboriginals. An indirect element is represented by cave paintings discovered in caves in the northern part of Queensland (Tresize, 1971; Vickers-Rich *et al.*, 1997) and in western Arnhem Land in the northern region of the Northern Territory and potentially depicting *Genyornis newtoni* (Gunn *et al.*, 2011). However, since no complete individual belonging to this species is known, it is hard to confirm that this is really a depiction of this bird. It is also impossible to conclude whether *Genyornis* was present in this region after its disappearance from the fossil record and coexisted with aboriginals, or if these paintings were drawn from the imagination and potentially derived from the oral transmission of the description of this bird, which would have then become extinct before these representations were painted. For now, these are the only representations we know and they remain at the center of numerous debates. A second element concerns the presence of traces of human tools on the bones of Dromornithidae discovered in the Penola site (Woods, 1866, 1882; Rich, 1979), which suggest that aboriginals may have hunted and eaten these birds. Unfortunately, these specimens have now disappeared and have never been illustrated, so that it is impossible to substantiate this observation.

The possible direct interactions between humans and these birds may have affected the populations of *Genyornis* especially if humans hunted them and collected their eggs (Murray & Vickers-Rich, 2004), thus progressively exterminating these birds which had already become endangered due to environmental changes. Besides the possibility of direct human action, humans have certainly had an indirect impact by substantially increasing fires in Australia, as shown by archeological, paleontological and botanical data (Murray & Vickers-Rich, 2004). Thus, 50000 years ago the areas covered by indigenous forest were so reduced that they were no longer enough to sustain the populations of *Genyornis*, whose habitat was thus progressively destroyed (Murray & Vickers-Rich, 2004).

The exact cause behind the extinction of Dromornithidae, whether environmental or anthropogenic, is still being debated (Miller *et al.*, 1999; Murray & Vickers-Rich, 2004). It is quite likely that significant climate changes, the alteration of the Australian flora and the competition with herbivorous marsupials endangered the populations of the last Dromornithidae, and that human action delivered the final blow to these already declining birds. However, the role and relative significance of these two factors are still being widely debated.

5

Phorusrhacidae

5.1. Historical background

Bravard (1860) seems to have been the first to mention the presence of giant birds in the Tertiary of South America in an autolithographic list of Argentine fossils, in which he mentioned "gigantic" birds in the sedimentary layers of the cliffs bordering the Paraná river in northern Argentina (Buffetaut, 2016a). However, it was at the end of the 1880s that discoveries made in Patagonia by both Carlos Ameghino and Francisco Moreno actually revealed to the scientific community the existence of Phorusrhacidae (Buffetaut, 2013d) thanks to the descriptions published by Moreno and Mercerat (1891) and Florentino Ameghino (1891, 1895). Andrews' publication on *Phororhacos inflatus* (1899), after the British Museum (Natural History) had bought Ameghino's collection of fossil birds, marked the internationalization of the studies on these birds. The discovery of remains of Phorusrhacidae by numerous paleontological expeditions in Patagonia and other regions of Argentina at the beginning of the 20th Century led to a certain number of publications about this group (e.g. Dolgopo De Saez, 1927; Sinclair & Farr, 1932; Patterson, 1941) as well as the discovery of fossils in Uruguay (Kraglievich, 1932) and Brazil (Alvarenga, 1982). Research on Phorusrhacidae is largely carried out in Argentina, the richest country in remains of these birds, and emphasis is often put on their paleobiology and paleoecology. Several reviews of the group have recently been published (Alverenga & Höfling, 2003; Agnolin, 2009; Alvarenga *et al.*, 2011; Tambussi & Degrange, 2013). The first description of a North American phorusrhacid specimen was published by Brodkorb (1963). The existence of Phorusrhacidae in the Eocene of Africa was reported by Mourer-Chauviré *et al.* (2011)), and fossil remains from the Eocene of Europe have been recently referred to this group (Angst *et al.*, 2013).

5.2. Taxonomy, systematics and classification

5.2.1. *Taxonomy*

The early stages of the nomenclatural history of Phorusrhacidae, in the 1890s, were quite chaotic, partly due to the rivalry between the Ameghino brothers and Moreno and Mercerat (Buffetaut, 2013d). Ameghino even considered for some time *Phorusrhacos* as a toothless mammal (1887, 1889). The number of taxa was highly overestimated, especially concerning the forms from the Miocene of Patagonia (Santa Cruz Formation), by the first authors who studied these birds (Moreno & Mercerat, 1891; Ameghino, 1895). A detailed history of the developments of phorusrhacid taxonomy is provided by Alvarenga & Höfling (2003). Dolgopol de Saez (1927) clarified an important point by clearly separating Brontornithidae from Phorusrhacidae, although this distinction (which is accepted in this book) is not currently accepted by all specialists (Alvarenga & Höfling, 2003; Alvarenga *et al.*, 2011).

While the number of taxa currently identified in Europe, Africa, and North America remains very low (only one for each of these continents, i.e. *Eleutherornis* in Europe, *Lavocatavis* in Africa and *Titanis* in North America), in line with the narrow stratigraphic range of this group in these regions, the situation is quite different in South America, where Phorusrhacidae diversified during the whole of the Tertiary, yielding a very abundant fossil record and consequently high taxonomic diversity. Many taxa described by Moreno and Mercerat (1891) and Ameghino (1891a, 1891b, 1895) have been synonymized; however, a large number of valid taxa have been erected since the beginning of the 20th Century. Despite some disagreements among recent authors (Alvarenga & Höfling, 2003; Agnolin, 2009; Alvarenga *et al.*, 2011) about the number of genera and species that should be distinguished and some differences in nomenclature (should the generic name *Tolmodus* Ameghino be preferred to *Patagornis* Moreno & Mercerat?), there is a relative consensus.

The issue of the appropriate spelling of the group name has been recently reviewed (Buffetaut, 2013c): the genus name *Phorusrhacos* Ameghino, 1887 should be used for the formation of higher-ranking taxa, Phorusrhacidae Ameghino, 1889 or possibly Phorusrhacoidea Ameghino, 1889 if such a superfamily is considered as justified. The terms "Phororhacidae" and "Phororhacoidea" are incorrect.

5.2.2. *Systematics*

The systematics of Phorusrhacidae has significantly evolved since the first discoveries at the end of the 19th Century as new taxa have been described. A first issue concerns the taxonomic rank of the group, as some authors (Patterson, 1941; Patterson & Kraglievich, 1960; Agnolin, 2009) have chosen to regard it as a superfamily (Phorusrhacoidea) that includes several families, whereas others (Alvarenga & Höfling, 2003; Alvarenga *et al.*, 2011; Degrange *et al.*, 2012; Tambussi & Degrange, 2013) prefer to consider only one family, Phorusrhacidae, including several subfamilies. The latter systematic arrangement seems to be favored by contemporary authors and is adopted here. Phylogenetic analyses of Phorusrhacidae have been provided especially by Agnolin (2009) and Alvarenga *et al.* (2011) – see also Degrange *et al.* (2015). Alvarenga and Höfling provided some detailed synonymic lists (2003).

Class Aves Linnaeus, 1758

Super-order Neognathae Pycraft, 1900

Order Cariamiformes Verheyen, 1957

Family Phorusrhacidae Ameghino, 1889

Sub-family Mesembriornithinae (Moreno, 1889)

Genus *Mesembriornis* Moreno, 1889

– *Mesembriornis incertus* (Rovereto, 1914)

Stratigraphic occurrence: Upper Miocene (Huayquerian) of Argentina.

– *Mesembriornis milneedwardsi* Moreno, 1889

Stratigraphic occurrence: Pliocene (Chapadmalalan) of Argentina.

Genus *Llallawavis* Degrange *et al.*, 2015

– *Llallawavis scagliai* Degrange *et al.*, 2015

Stratigraphic occurrence: Pliocene (Chapadmalalan) of Argentina.

Genus *Procariama* Rovereto, 1914

– *Procariama simplex* Rovereto, 1914

Stratigraphic occurrence: Upper Miocene (Huayquerian) of Argentina.

Sub-family Psilopterinae Dolgopol de Saez, 1927

Genus *Psilopterus* Moreno & Mercerat, 1891

– *Psilopterus affinis* (Ameghino, 1899)

Stratigraphic occurrence: Upper Oligocene (Deseadan) of Argentina.

– *Psilopterus bachmanni* (Moreno & Mercerat, 1891)

Stratigraphic occurrence: Lower Miocene (Santacrucian) of Argentina.

– *Psilopterus colzecus* Tonni and Tambussi, 1988

Stratigraphic occurrence: Upper Miocene (Chasicoan) of Argentina.

– *Psilopterus lemoinei* (Moreno & Mercerat, 1891)

Stratigraphic occurrence: Lower Miocene (Santacrucian) of Argentina.

Sub-family Phorusrhacinae Ameghino, 1889

Genus *Kelenken* Bertelli *et al.*, 2007

– *Kelenken guillermoi* Bertelli *et al.*, 2007

Stratigraphic occurrence: Middle Miocene (Colloncuran) of Argentina.

Genus *Devincenzia* Kraglievich, 1932

– *Devincenzia pozzi* (Kraglievich, 1931)

Stratigraphic occurrence: Upper Miocene/Pliocene of Argentina.

Genus *Titanis* Brodkorb, 1963

– *Titanis walleri* Brodkorb, 1963

Stratigraphic occurrence: Pliocene of the United States.

Genus *Paraphysornis* Alvarenga, 1993

– *Paraphysornis brasiliensis* (Alvarenga, 1982)

Stratigraphic occurrence: Upper Oligocene (Deseadan) of Brazil.

Genus *Andrewsornis* Patterson, 1941

– *Andrewsornis abbotti* Patterson, 1941

Stratigraphic occurrence: Upper Oligocene (Deseadan) of Argentina.

Genus *Andalgalornis* Patterson & Kraglievich, 1960

– *Andalgalornis steulleti* (Kraglievich, 1931)

Stratigraphic occurrence: Upper Miocene (Huayquerian) of Argentina.

Genus *Patagornis* Moreno & Mercerat, 1891

– *Patagornis marshi* Moreno & Mercerat, 1891

Stratigraphic occurrence: Lower Miocene (Santacrucian) of Argentina.

Genus *Phorushacos* Ameghino, 1887

– *Phorusrhacos longissimus* Ameghino, 1887

Stratigraphic occurrence: Lower Miocene (Santacrucian) of Argentina.

Genus *Physornis* Ameghino, 1895

– *Physornis fortis* Ameghino, 1895

Stratigraphic occurrence: Upper Oligocene (Deseadan) of Argentina.

Phorusrhacidae *incertae sedis*

Genus *Eleutherornis* Schaub, 1940

– *Eleutherornis cotei* (Gaillard, 1936)

Stratigraphic occurrence: Middle Eocene (Lutetian) of France and Switzerland.

Genus *Lavocatavis* Mourer-Chauviré *et al.*, 2011

– *Lavocatavis africana* Mourer-Chauviré *et al.*, 2011

Stratigraphic occurrence: Lower Eocene of Central Algeria

5.2.3. *Classification*

Moreno and Mercerat (1891) initially placed Phorusrhacidae into a separate order, Stereornithes, a classification accepted by several authors in the last decade of the 19th Century (Ameghino, 1895; Lydekker, 1896). Moreno and Mercerat (1891) attributed to them features shared by Herodiones (herons and storks), Accipitres (raptors) and Anseres (ducks) and considered them as an order transitional between Vulturidae and Anatidae, while also identifying some reptilian characters in them. Ameghino, after initially associating them with ratites (Ameghino, 1981), considered them as intermediate between Accipitres and Anseres, consequently classifying them among carinates (Ameghino, 1895). Lydekker (1896) still hesitated to classify them among either ratites or carinates. Andrews (1899) seems to have been the first to note significant similarities between *Phorusrhacos* and seriemas (*Cariama* and *Chunga*), which he considered aberrant Gruiformes.

In 1927, Dolgopol de Saez was still using the term Stereornithes, in which, however, she placed only Phorusrhacidae, as Brontornithidae were placed in a different order, Brontornithes. The work carried out by Dolgopol de Saez is particularly significant, as it led to a clear distinction between Brontornithidae and Phorusrhacidae. This viewpoint, adopted among others by Agnolin (2007, 2013) and several other authors (Degrange *et al.*, 2012; Tambussi & Degrange, 2013), is accepted in this work. Kraglievich (1932) generally followed the suggestions made by Dolgopoel de Saez by distinguishing Stereornithes from Brontornithes. Lambrecht (1933) classified Stereornithes (including Phorusrhacidae and Brontornithidae) within the large order of Telmatoformes, which also included Cariamae, Gruae, Limicolae and Laro-Limicolae. This classification is by now only relevant from a historical perspective, but it had the advantage of relating the Phorusrhacidae to some extant groups, especially Cariamae (seriemas), following Andrews in this respect.

Later authors have generally accepted that Phorusrhacidae and seriemas are closely related, while also rejecting the term Stereornithes. Patterson and Kriegllevich (1960), for example, placed "Phororhacoidea" besides Cariamoidea in the sub-order Cariamae, which they placed in the order Grues. Brodkorb (1967) classified Phorusrhacidae with Cariamidae within the sub-order Cariamae, which he placed in the order Ralliformes.

The most recent classifications based on phylogenetic analyses reach similar conclusions, underlining the proximity between Phorusrhacidae and seriemas, which are all classed among Cariamae, which also include several extinct forms that do not belong to Phorusrhacidae, such as the European taxa *Elaphrocnemus* and *Strygogyps* (Agnolin, 2009; Alvarenga *et al.*, 2011; Figure 5.1).

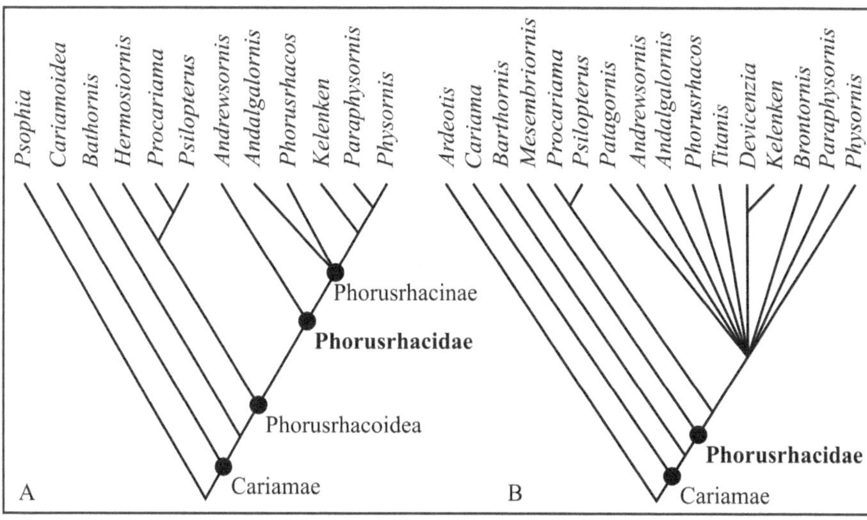

Figure 5.1. *Phylogenetic position of Phorusrhacidae following the hypotheses of Agnolin (2009, A) and Alvarenga* et al. *(2011, B)*

5.3. Stratigraphic and geographical distribution

Phorusrhacidae have long been known exclusively from South America. We now know that they were also present in North America, Africa and Europe (Figure 5.2). While it is clear that North American Phorusrhacidae are recent immigrants that came from South America during the Great American Faunal Interchange, which began in the Pliocene, the biogeographical relationships between South American, African and European forms in the Paleogene are less clear. The geographical distribution of Eocene Phorusrhacidae apparently implies dispersal across stretches of ocean (Angst *et al.*, 2013).

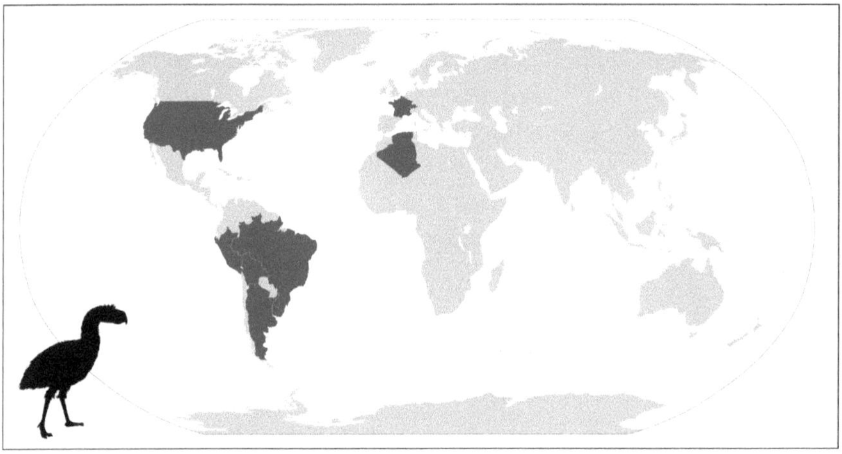

Figure 5.2. *Geographical distribution of the countries where body fossils of Phorusrhacidae have been discovered. For a color version of the figure, see www.iste.co.uk/angst/birds.zip*

5.3.1. *South America*

5.3.1.1. *Paleocene*

Limb bones from the karst fossil sites of Itaborai (in the state of Rio de Janeiro, Brazil, Figure 5.3) and dating back to the Upper Paleocene have been attributed to Phorusrhacidae and described as *Paleopsilopterus itaboraiensis* by Alvarenga (1985). However, Agnolin (2009) noted different features that separate this taxon from Phorusrhacidae and suggested that it should be classed among Idiomithidae. Tambussi and Degrange (2013) thought that it belongs to an indeterminate family of Cariamiformes.

5.3.1.2. *Eocene*

The oldest well-documented remains of Phorusrhacidae in South America include the distal end of a tibiotarsus attributed to a psilopterine described by Tonni & Tambussi (1986) from the Casamayoran (Upper Eocene), (Argentina, Figure 5.3). However, slightly older Phorusrhacidae have been identified but not described in detail by Sedor *et al.* (2016) from the Guabirotuba Formation in the state of Parana (southern Brazil). They belong to a fauna referred to the Barrancan (end of the Middle Eocene).

The fragmentary remains attributed to Psilopterinae by Acosta Hospitaleche and Tambussi (2005), from the Mustersan (Terminal Eocene) of Chubut, are placed among Idiornithidae by Agnolin (2009) and considered of dubious affinities by Tambussi and Degrange (2013).

5.3.1.3. *Oligocene*

The Deseadan (Late Oligocene) of the Argentinian part of Patagonia has yielded several – mostly relatively incomplete – remains of Phorusrhacidae, including small forms (Psilopterinae; Buffetaut, 2014a) and other much larger forms among which are *Physornis fortis*, known from very fragmentary remains (Ameghino, 1895), and *Andrewsornis abbotti*, represented by a relatively complete skull (Patterson, 1941).

The Tremembé Formation (Deseadan, Late Oligocene, Figure 5.3.) of the Taubaté basin (in the state of São Paulo, Brazil), has yielded an incomplete skeleton of a very large phorusrhacid, described as *Paraphysornis brasiliensis* by Alvarenga (1982, 1993).

The Salla Beds (Figure 5.3), of Deseadan age, in the Salla-Luribay basin of Bolivia have yielded fragmentary remains that may be attributed to Phorusrhacidae, among which are a small form (Psilopterinae) and another very large form that resembles *Paraphysornis* (Buffetaut, in prep.).

In Peru, the Moquegua Formation (Deseadan, Figure 5.3) has yielded an ungual phalanx that may be referred to an indeterminate form of Phorusrhacidae (Shockey *et al.*, 2009).

5.3.1.4. *Miocene and Mio-Pliocene*

Phorusrhacidae from the Colhuehuapian (Early Miocene, Figure 5.3) of the Argentinian part of Patagonia are only known from very succinct remarks ("*Phororhacos chubutensis*") by Ameghino (1901) in faunal lists about the "*Colpodon* Beds". Because of the lack of descriptions, it is difficult to assess the status of these taxa.

The Santa Cruz Formation (Santacrucian, late Early Miocene, Figure 5.3) in southern Patagonia (provinces of Santa Cruz and Chubut, Argentina) has yielded a diverse assemblage of Phorusrhacidae of different sizes (Degrange *et al.*, 2012), including Psilopterinae (*Psilopterus lemoinei*, *Psilopterus bachmanni*) and Phorusrhacinae (*Patagornis marshi*, *Phorusrhacos longissimus*).

The remains of *Kelenken guillermoi*, considered as the largest of all phorusrhacids (Bertelli *et al.*, 2007), have been found in the Collón Curá Formation (Colloncuran, Early Miocene, Figure 5.3) of the province of Rio Negro (Argentina).

The Cerro Azul Formation (Huayaquerian, Late Miocene, Figure 5.3) of the provinces of La Pampa and Buenos Aires has yielded remains of a mesembriornithine, *Procariama simplex* (Vezzosi, 2012a, 2012b).

The Puerto Madryn Formation of Chubut province (Argentinian part of Patagonia, Figure 5.3), deposited in a shallow coastal environment, is thought to date back to the Late Miocene. It has mostly yielded remains of marine birds and also a few bones belonging to a psilopterine (Tambussi & Degrande, 2013).

The only phorusrhacid known from the Chasicó Formation (Chasicoan, Late Miocene, Figure 5.3) of Buenos Aires province is *Psilopterus colzecus*, which belongs to Psilopterinae (Tonni & Tambussi, 1988).

The Ituzaingó Formation (Upper Miocene, Figure 5.3), which belongs to the "Mesopotamian" of the province of Entre Rios (northern Argentina), has yielded remains of "gigantic" birds since the middle of the 19th Century (Bravard, 1860; Buffetaut, 2016a). The remains of Phorusrhacidae found there are generally fragmentary. Noriega and Agnolin (2008), as well as Diederle and Noriega (2013), reported specimens that may be referred to *Devincenzia pozzi* and *Andalgalornis steulleti* as well as indeterminate

Phorusrhacidae. It should be noted that the type of *Devincenzia pozzi*, described by Kraglievich (1932), comes from a formation in Uruguay that probably dates back to the same period. A skull associated with this large species, initially described as *Onactornis depressus* by Cabrera (1939), has been found in the province of Buenos Aires and is thought to be Pliocene in age.

The Andalhualá Formation (Huayaquerian, Late Miocene to Early Pliocene, Figure 5.3) in north-western Argentina (Catamarca province) has yielded remains of three species of Phorusrhacidae: *Andalgalornis steullti*, *Mesembriornis incertus* and *Procariama simplex*. Tambussi and Degrange (2013), however, pointed out that the exact stratigraphic origin of some specimens collected in the past (Rovereto, 1914) is not always known with precision.

5.3.1.5. *Pliocene*

The Chapadmalal Formation (Chapadmalalan, Pliocene, Figure 5.3) in Buenos Aires province has yielded the large mesembriornithine *Mesembriornis milneedwardsi* as well as the remarkably well-preserved skeleton of another mesembriornithine, *Llallawavis scagliai* (Degrange *et al.*, 2015).

5.3.1.6. *Pleistocene*

The issue of the persistence of Phorusrhacidae in South America until the Pleistocene is of utmost importance. A large tibiotarsus has been reported from the Raigón Formation of Uruguay by Tambussi *et al.* (1999), but it may be Pliocene in age (Agnolin, 2013). Several rather fragmentary remains from the Dolores Formation (Middle to Upper Pleistocene, Figure 5.3), also in Uruguay, have been described by Alvarenga *et al.* (2010) and Jones *et al.* (2016). The attribution of the first remains discovered to Phorusrhacidae has been contested, but if their identification is correct, they constitute the most recent remains of Phorusrhacidae hitherto found (see section 5.6, "Extinction").

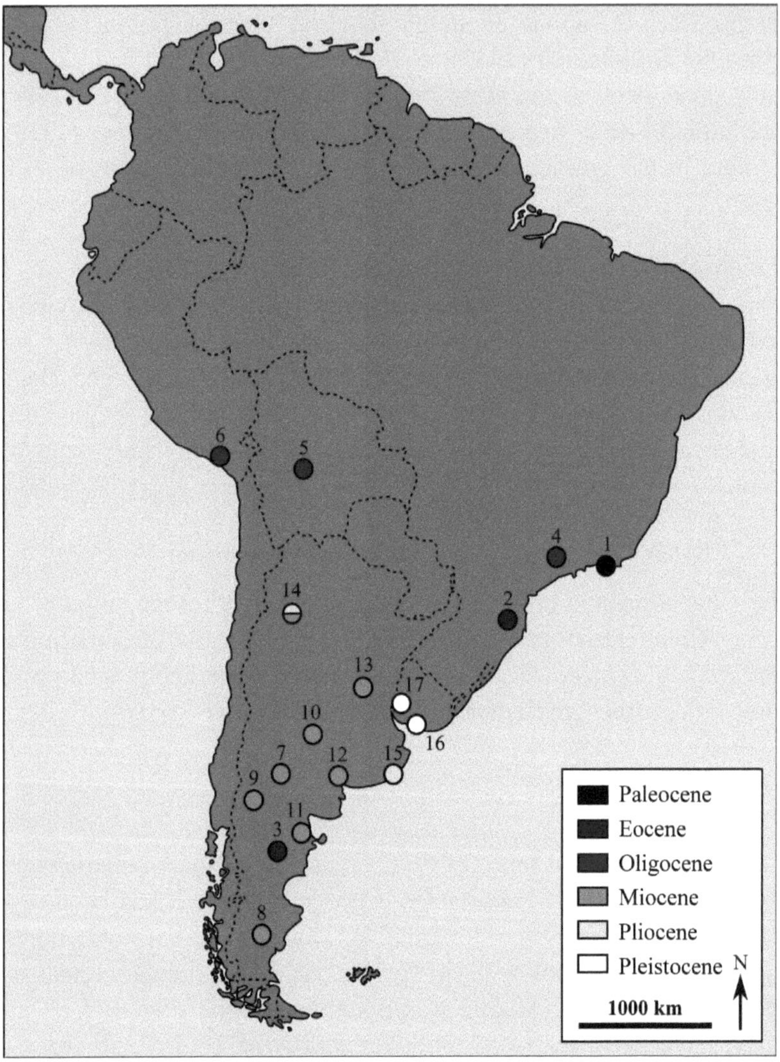

Figure 5.3. *Geographical distribution of the main fossil sites of Phorusrhacidae in South America. Paleocene site: 1 – Itaboraí Formation; Eocene sites: 2 – Guabirotuba Formation, 3 – Casamayorian in Chubut province; Oligocene sites: 4 – Tremembé Formation,, 5 – Salla Beds, 6 – Moquegua Formation; Miocene sites: 7 – Colhuehuapian of the Argentinian part of Patagonia, 8 – Santa Cruz Formation, 9 – Collón Curá Formation, 10 – Cerro Azul Formation, 11 – Puerto Madryin Formation, 12 – Chasicó Formation, 13 – Ituzaingó Formation; Mio-Pliocene site: 14 – Andalhualá Formation; Pliocene site: 15 – Chapadmalal Formation; Pleistocene sites: 16 – Raigón Formation, 17 – Dolores Formation*

5.3.2. *North America*

5.3.2.1. *Pliocene*

The presence of Phorusrhacidae in North America was reported for the first time by Brodkorb (1963), who described the distal end of a tarsometatarsus and a phalanx from Florida as *Titanis walleri*. It was initially suggested that they were Late Pleistocene in age. However, in their review of *Titanis walleri*, which describes additional specimens from Florida as well as from Texas (Figure 5.4), Gould and Quitmyer (2005) concluded that all the material from North America is probably Pliocene in age.

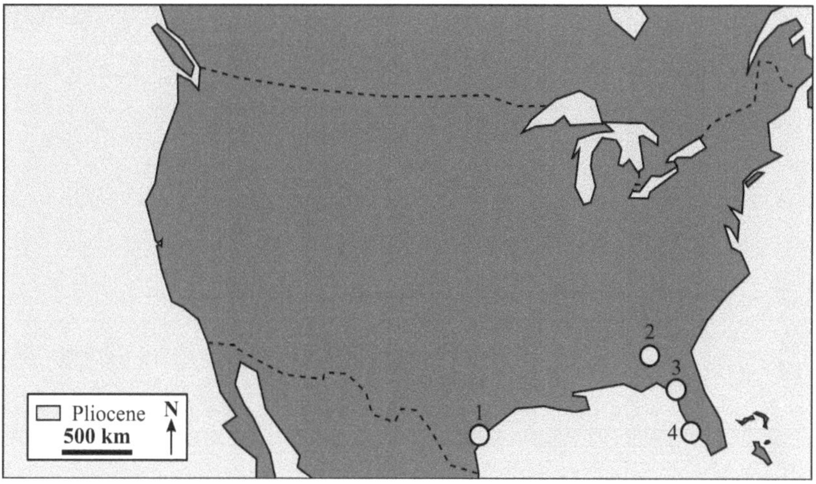

Figure 5.4. *Geographical distribution of the main fossil sites of Phorusrhacidae in North America. Pliocene site in Texas: 1 – Nueces River; Pliocene sites in Florida: 2 – Santa Fe River, 3 – Inglis, 4 – Port Charlotte*

5.3.3. *Africa*

5.3.3.1. *Eocene*

Only one specimen that may be attributed to Phorusrhacidae has been identified from Africa. It is a femur found in Algeria in the Glib Zegdou Formation (Figure 5.5) and referred to the end of the Early Eocene or the beginning of the Middle Eocene. The fossil was described as *Lavocatavis africana* by Mourer-Chauviré *et al.* (2011) and regarded as "Phorusrhacoidea indet".

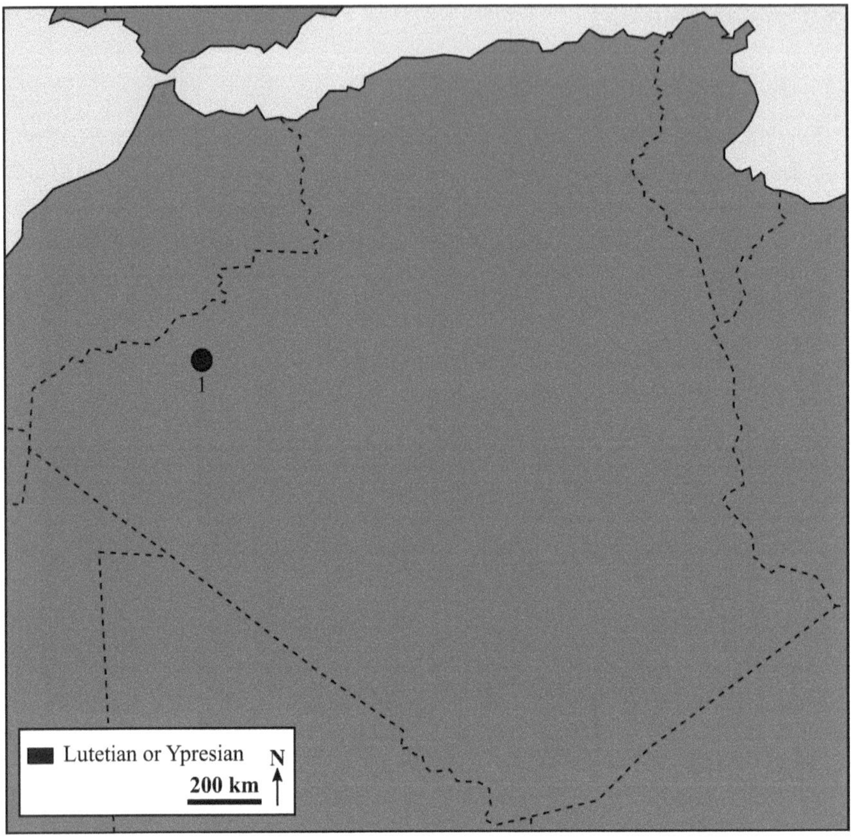

Figure 5.5. *Position of the fossil site of Phorusrhacidae in North Africa. Lutetian or Ypresian site: 1 – Glib Zegdou Formation*

5.3.4. *Europe*

5.3.4.1. *Eocene*

Gaillard (1936, 1937) described as *Diatryma* (?) *cotei* tarsometatarsus fragments and phalanges from the fossil site of Lissieu, close to Lyon (France), which is Late Lutetian in age (Middle Eocene, MP 14, Figure 5.6). A few years later, Schaub (1940) attributed to a ratite, which he named *Eleutherornis helveticus*, an incomplete pelvis from a karst pocket in Egerkingen (Switzerland, Figure 5.6), also Late Lutetian in age. Angst *et al.* (2013) showed that all this material, both French and Swiss, can be attributed to the same taxon, which must be called *Eleutherornis cotei* and belongs to the Phorusrhacidae. This middle-sized bird was referred to Psilopterinae, but the limited material available makes it difficult to confirm this attribution.

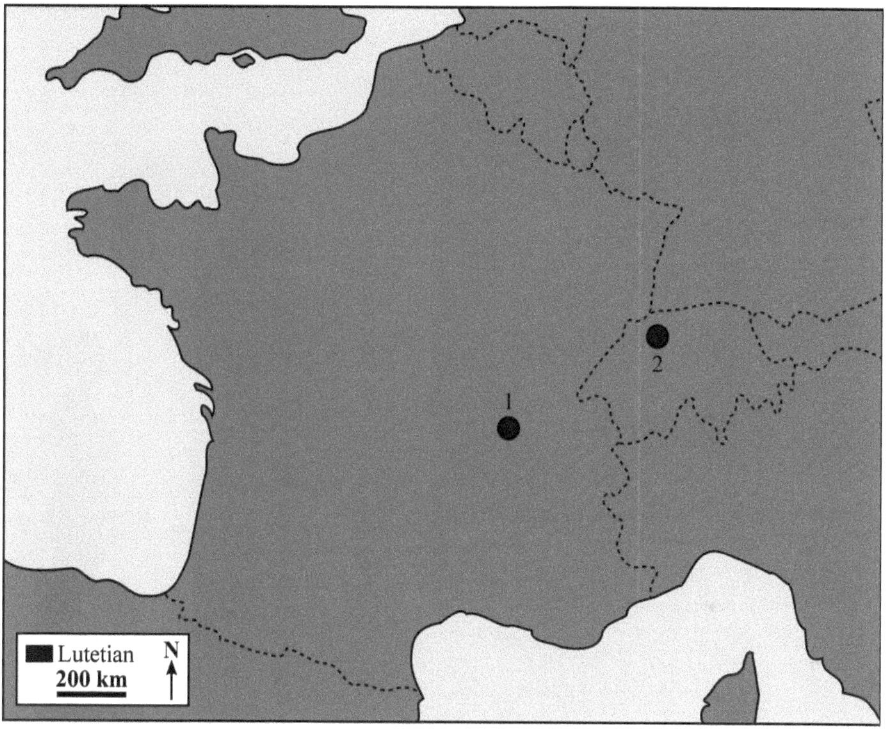

Figure 5.6. *Geographical distribution of the main fossil sites of Phorusrhacidae in Europe. Lutetian sites: 1 – Lissieu, 2 – Egerkingen*

5.4. Anatomy

Despite the significant differences in size between the various taxa (between 70 and 240 cm in height, head raised, Figurc 5.7), Phorusrhacidae are relatively anatomically homogenous.

Figure 5.7. *Skeleton and restoration of* Phorusrhacos longissimus

The skull is known from a fairly large number of specimens. It is remarkable because of its large dimensions in relation to the rest of the body and of its strong beak, which is much compressed laterally (Figure 5.8). The tip of the beak is markedly hooked downward, as in birds of prey, thus providing evidence of the carnivorous diet of these birds. A significant point is that the skull is not kinetic, but rigid; thus, the beak had lost its mobility in relation to the rear of the skull, unlike the condition in most birds. The mandible is elongated, with a long and generally narrow symphysis. Its posterior rami are quite low, and a large external mandibular fenestra runs through them. The neck is relatively long; however, the cervical vertebrae are not particularly elongate.

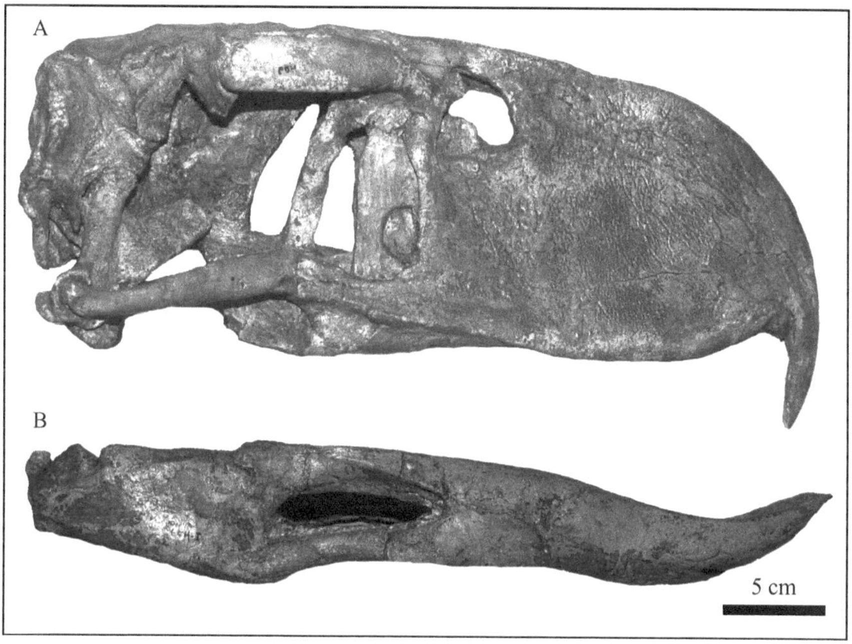

Figure 5.8. *Cast of the skull (A) and mandible (B) of* Andalgalornis steulleti *(specimen n° FM P14357). Photo: Eric Buffetaut. For a color version of the figure, see www.iste.co.uk/angst/birds.zip*

The pelvis is elongated and highly compressed laterally, in line with the cursorial type of locomotion of these birds (Figure 5.9). The front limb is reduced and especially so in large individuals. It is less reduced in Psilopterinae which, according to some authors, might still have been able to fly clumsily (see below). The hind limb is very long, and the most elongated segment is the tibiotarsus, followed by the tarsometatarsus (Figure 5.9). The latter bone is generally long and narrow, but it is more robust in some particularly large and robust forms, with implications for locomotion (see section "locomotion"). The ungual phalanges are significantly hooked into claws and laterally compressed (Figure 5.9).

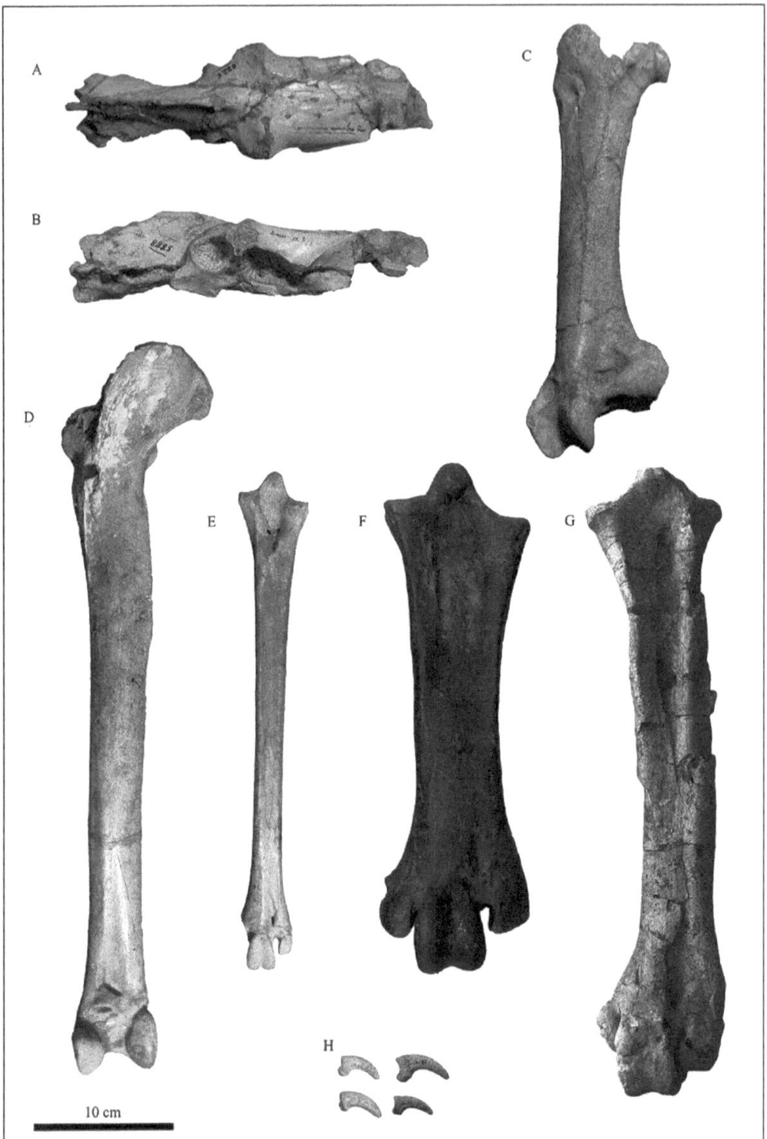

Figure 5.9. *Pelvis of* Procariama simplex *(MACN 8225) seen dorsally (A) and laterally (B); pelvic limb bones of* Hermosiornis milneedwardsi *(MACN 5944) including a femur (C), a tibiotarsus (D) and a tarsometatarsus (E) (Photos Eric Buffetaut); Tarsometatarsus of* Paraphysornis brasiliensis *(F) (photo Delphine Angst), Tarsometatarsus of* Kelenken guillermoi, *BAR 3877-11 (G) (after Bertelli* et al., *2007), terminal phalanges of* Eleutherornis cotei *(H) (after Angst* et al., *2013). For a color version of the figure, see www.iste.co.uk/angst/birds.zip*

5.5. Paleophysiology and paleoecology

5.5.1. *Size and body mass*

The large size of some Phorusrhacidae has been emphasized since the first relevant studies, as authors often insist on the fact that the skull of some forms was larger than that of a horse. Nonetheless, there is a wide range of sizes within this group: Psilopterinae are the smallest forms, whereas some Phorusrhacinae reached a very large size (Table 5.1).

	Height measured from the back (cm)	Height measured from the head (raised) (cm)
Psilopterus bachmanni	60	80
Procariama simplex	70	
Andalgalornis steulleti	90–100	
Patagornis marshi	90–100	
Mesembriornis milneedwardsi	110–120	170
Phorusrhacos longissimus	140	240
Paraphysornis brasiliensis	140	240

Table 5.1. *Summary of the different sizes estimated for the several groups of Phorusrhacidae. Values obtained by Alverenga and Höfling (2003)*

The first body mass estimates for Phorusrhacidae were put forward by Alvarenga and Höfling (2003), who apparently relied on comparisons with present-day birds (*Struthio*, *Cariama*, etc.) of similar sizes (Table 5.2). Afterward, Degrange *et al.* (2012) estimated the weight of several Phorusrhacidae from the Santa Cruz Formation (Lower Miocene) based on the method of Campbell and Marcus (1992) (Table 5.2).

Method	Campbell & Marcus (1992) (femur) (kg)	Campbell & Marcus (1992) (tibiotarsus) (kg)	Other (kg)
Psilopterus bachmanni	4.5 [1]		5 [2]
Procariama simplex			10 [2]
Psilopterus lemoinei	8 [1]	13 [1]	
Patagornis marshi	26 [1]	34 [1]	45–50 [2]
Andalgalornis steulleti			45–50 [2]
Mesembriornis milneedwardsi			70 [2]
Phorusrhacos longissimus	94 [1]	153 [1]	130[2]
Paraphysornis brasiliensis			180 [2]

Table 5.2. *Summary of the body mass values estimated for the different groups of Phorusrhacidae. [Bibliographical reference]: [1] Degrange* et al. *(2012), [2]: Alvarenga and Höfling (2003)*

5.2.2. *Locomotion*

Characterized by substantially elongated hindlimbs, especially their tibiotarsus and tarsometatarsus, the Phorusrhacidae have generally been regarded as running birds able to reach high speeds while chasing their prey. Nevertheless, there are significant differences between the various taxa in the robustness of the skeleton of the hindlimbs in relation with body mass. Angst *et al.* (2016) assessed the type of locomotion of different Phorusrhacidae by using a method that relies on the proportions (length/width ratio) of the tarsometatarsus. Their results, supplemented by those obtained by Angst and Chinsamy (in press), indicate a cursorial type of locomotion (length/width ratio for the tarsometatarsus greater than 12) for *Psilopterus lemoinei*, *Psilopterus bachmanni* and *Phorusrhacos longissimus* (Figure 5.10). On the contrary, *Paraphysornis brasiliensis* and *Kelenken guillermoi* figure among the graviportal birds (length/width ratio for the tarsometatarsus greater than 12, Figure 5.10), which is probably associated with their general robustness and considerable body mass. Therefore, it can be assumed that *Paraphysornis* and *Kelenken* had a slower type of locomotion than that of smaller Phorusrhacidae of a lighter build, affecting their feeding mode. These authors (Angst *et al.*, 2016; Angst & Chinsamy, in press) mentioned the possibility of a diet based on scavenging or ambush hunting. We may also suppose that these large and massive birds attacked large and fairly slow prey, such as large mammals (e.g. the Astrapotheria and Pyrotheria found with *Paraphysornis brasiliensis* in the Tremembé Formation or giant sloths).

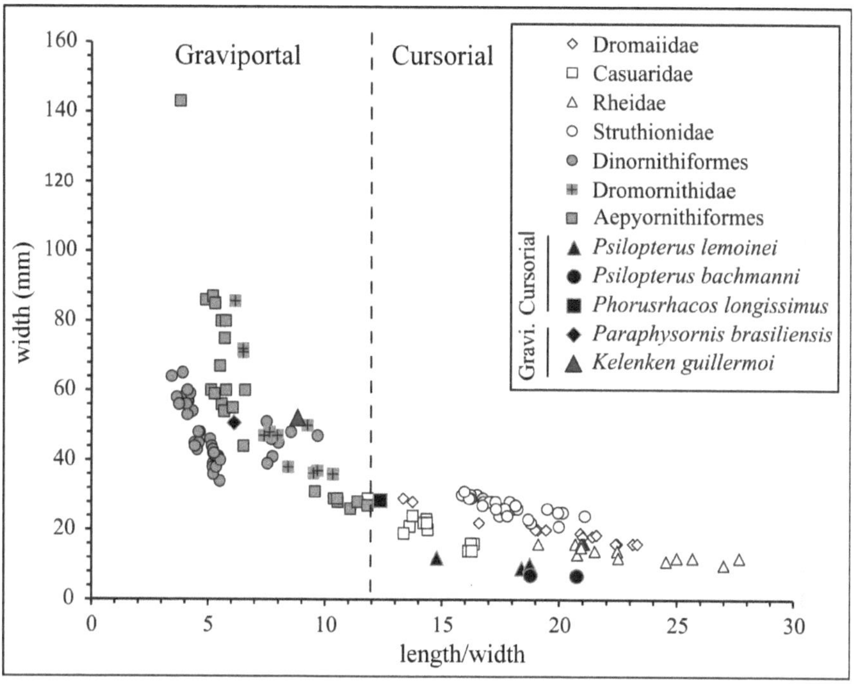

Figure 5.10. *Estimate of the type of locomotion of Phorusrhacidae based on the method used by Angst* et al. *(2016)*

Blanco and Jones (2005) attempted to assess the speed of some Phorusrhacidae by using a method that relies on the resistance of the tibiotarsus. Their results indicate that *Patagornis marshi* and a large Phorusrhacinae (estimated weight: 350 kg – which seems excessive) would have reached a speed of about 50 km/h, whereas *Mesembriornis milneedwardsi* would have reached 97 km/h. The latter speed seems quite high, and the method should be used with caution. As the authors acknowledge, the speed (112 km/h) estimated by following this method for the small graviportal moa *Anomalopteryx didiformis* is incredibly excessive, showing that the resistance of the tibiotarsus is not linked merely to the speed of locomotion.

It should be pointed out that no footprints clearly attributable to Phorusrhacidae, which may provide some information about the locomotion of these birds, are known. Tridactyl footprints left by large land birds from the Paleocene-Eocene of King George Island (South Shetland Islands) have been attributed either to ratites or Phorusrhacidae (Covacevich & Rich, 1982). Cenizo (2012) tended to favor the first solution, especially because reports of phorusrhacid bones from Antarctic regions are apparently erroneous, whereas ratites are known from that area.

5.3.3. *Flight abilities*

It is generally acknowledged that middle-sized or large Phorusrhacidae could not fly because of their weight and reduced wings. The issue of the ability to fly only arises for certain Psilopterinae, especially the species belonging to the genus *Psilopterus*. With a weight ranging between 5 and 10 kg (Degrange & Tambussi, 2011; Degrange *et al.*, 2015), these forms were not too heavy to be able to fly – several extant larger birds can fly (the heaviest extant bird able to fly, i.e. the kori bustard (*Ardeotis kori*), can weigh as much as 19 kg, and we know fossils of flying birds that were undoubtedly even heavier, especially some Teratornithidae and Pelagornithidae). Nevertheless, the wings of Psilopterinae are shorter than the hindlimbs, and it is likely that these birds moved around more by walking or running than by flying, suggesting a type of locomotion similar to that of Otididae and Cariamidae (Degrange *et al.*, 2015). If they were actually able to fly, their flight must have been rather clumsy, like that of present-day seriemas (Cariamidae). However, whether Psilopterinae could fly or not remains a debated topic. Some authors think that it may have been possible (Tonni & Tambussi, 1988; Acosta Hospitaleche & Tambussi, 2005), whereas others regard it as too unlikely when limb proportions are taken into consideration (Alvarenga & Höfling, 2003). Degrange *et al.* (2015) concluded from a study on the front limb of *Psilopterus bachmanni*, a small form, that this bird was probably able to fly awkwardly.

On the basis of the morphology of the carpometacarpus, supporting a hemispheric articulation surface, Chandler (1994) supposed that the wing of

Titanis walleri (and related species) was robust and characterized by a moveable thumb that was probably equipped with a powerful claw used by the bird to hold its prey. This hypothesis has given rise to several restorations where *Titanis* is represented with a sort of clawed digit on its wings. Gould and Quitmyer (2005) contested Chandler's interpretation and pointed out that the articulation in question is similar to what may be observed in Cariamidae. Most importantly, the wing of *Titanis* is particularly reduced, especially when compared to the wings of Psilopterinae. Thus, there is no reason to suppose that *Titanis* had powerful wings with highly moveable thumbs equipped with a claw, and the wings of Phorusrhacidae certainly could not help them hold their prey in any way.

5.5.4. *Hearing*

The remarkably well-preserved skeleton of *Llallawavis scagliai* described by Degrange *et al.* (2015) allowed these authors to obtain some interesting information about the hearing of this Pliocene mesembriornithine. The study of the inner ear shows an average auditory sensitivity (2300 Hz), which is lower than the average obtained for present-day species of Cariamiformes. *Llallawavis scagliai* would have had good auditory abilities at low frequencies for intraspecies communication or hunting.

5.5.5. *Diet*

Since the discovery of sufficiently complete cranial remains, it has been evident that phorusrhacids were carnivorous birds (Ameghino, 1889), as shown by their high and narrow beak, whose extremity was strongly hooked and actually similar to that of present-day birds of prey and different from the tall but unhooked beaks of Gastornithidae and Dromornithidae. It should also be noted that the lower jaw of Phorusrhacidae, characterized by an elongate and large external mandibular fenestra, does not show any vast surface for the insertion of the mandibular adductor muscles. As Angst *et al.* (2014a) showed, these adductors, which are very developed in herbivorous birds, are weak in carnivorous forms. The reduced attachment areas of these muscles in Phorusrhacidae, therefore, are in keeping with a carnivorous diet. The ungual phalanges, which are compressed laterally and very hooked, with a strong flexor tubercle, show that the foot had strong claws comparable to the talons of raptors. The nickname "terror birds", which is often used to refer to them, conjures up their image of formidable predators, as is shown in the many restorations that depict them while preying on mammals of different sizes. However, given their widely different sizes, it is clear that the several species of Phorusrhacidae did not attack the same prey; small Psilopterinae certainly preyed on modest-sized vertebrates and even invertebrates, whereas large forms could feed on much larger animals. It is interesting to note that in most formations containing remains of Phorusrhacidae, species of different sizes occur, from small Psilopterinae to very large Phorusrhacinae. Therefore, it seems very likely that these different species shared distinct carnivorous ecological niches. A scavenging behavior has occasionally been suggested (Tonni & Noriega, 1998), but it seems difficult to demonstrate.

The evolutionary radiation of Phorusrhacidae in South America during the Tertiary is probably associated with the existence on this isolated continent of specific ecosystems, where non-avian land carnivores were marsupials and land crocodiles, which were probably slow-moving, allowing large carnivorous cursorial birds to develop and occupy specific ecological niches, especially in the open environments that expanded during the climate changes of the Neogene (Tambussi & Degrange, 2013).

Degrange *et al.* (2012) assessed the bite force of several Phorusrhacidae from the Santa Cruz Formation, extrapolated on the basis of their body mass, which was in turn assessed on the basis of the circumference of the femur. Logically, Psilopterinae show weaker bite force (49.5 N for *Psilopterus bachmanni* and 64 N for *Psilopterus lemoinei*), whereas higher values are obtained for *Patagornis marshi* (110 N) and *Phorusrhacos longissimus* (196 N).

Degrange *et al.* (2010) also applied finite element analysis to a skull of *Andalgalornis steulleti* to reconstruct the diet of this middle-sized species of Phorusrhacidae (estimated body mass ~ 40 kg). A significant point is the loss of cranial kinesis through the fusion of bone elements and the stiffening of flexion areas. Like *Andalgalornis*, all middle-sized to large Phorusrhacidae apparently also had an akinetic and rigid skull. Bite force has been estimated to be 133 N at the tip of the beak. The reconstruction of the forces exerted on the beak, compared with the results obtained for a seriema (a close relative of Phorusrhacidae) and an eagle (a large carnivorous bird), suggests that the beak of *Andalgalornis* was particularly constrained laterally, whereas dorsoventral constraints were less marked. The authors conclude that this bird could hardly grab and hold large and writhing preys with its beak. Consequently, either it fed mainly on fairly small prey, which it could swallow whole, or it attacked larger prey by using precise and repeated sagittal strikes, stabbing them from above with the hooked point of its beak, or its diet was based on scavenging, as has been proposed for other Phorusrhacidae (Angst & Chinsamy, 2016), feeding then on dead prey that ran no risk of squirming.

In a popular science article about Phorusrhacidae, Trouessart (1895, p. 91) briefly reported on something that may constitute an additional argument in favor of the carnivorous diet of Phorusrhacidae, viz. the existence of fossilized pellets. These pellets are "made of long bones belonging to large animals crushed and with the skull in the center", quite similar (apart from the larger size) to those regurgitated by nocturnal birds of prey. The information conveyed by Trouessart comes from a letter from Ameghino (transcribed by Torcelli, 1936, p. 573), where he claimed to possess "petrified masses preserving skeletons of large rodents, *Protypotheridae* and even *Proterotheridae*, with all their bones crushed and corroded, piled on with no apparent order and forming a nearly spherical mass with the skull in the center". Nevertheless, Ameghino does not seem to have ever mentioned or figured these objects in any of his publications, so that it is difficult to decide whether they were actually regurgitated pellets that may be attributed to Phorusrhacidae.

Ameghino's interpretation may be backed up by the study carried out by Nasif *et al.* (2009), who described two regurgitated pellets from the Andalhualá Formation (Upper Miocene) in Catamarca province (northern Argentina). These are calcified masses a few centimeters long and containing bones and teeth of small rodents, very similar to the regurgitated pellets of present-day raptors. The teeth show traces of corrosion attributed to gastric juices. As the only carnivorous birds recorded from the Andalhualá Formation are Phorusrhacidae, Nasif *et al.* (2009) supposed that these pellets may be attributed to the smallest of the species identified in these layers, i.e. *Procariama simplex*, which reached 70 cm in height for an estimated weight of 10 kg, as there is a relationship between the size of the pellets and that of the bird that produces them. Considering that present-day seriemas, which are closely related to Phorusrhacidae, regurgitate pellets, this interpretation seems plausible and may provide useful information about the diet of some Phorusrhacidae.

According to Tambussi (2011), Phorusrhacidae must have been able to swallow their small prey whole and rip apart the larger ones with their claws.

5.5.6. *Reproduction*

No eggs of Phorusrhacidae are known. In a letter sent to Trouessart in 1895 (Torcelli, 1936, p. 573), Ameghino was surprised about the absence of egg remains in the Santa Cruz Formation of Patagonia, which, however, contains many skeletal remains of these birds, and found it reasonable to draw the conclusion that the eggs of Phorusrhacidae had "no calcareous shell", like those of some reptiles. This hypothesis is obviously highly unlikely.

5.7. Paleoenvironment

The environments inhabited by Phorusrhacidae have been analyzed with more or less precision according to the places and geological periods considered. As for the European Eocene sites (Lissieu and Egerkingen), it may be mentioned that they are from karst deposits, suggesting that *Eleutherornis* lived on calcareous plateaus with relatively few bodies of water on the surface.

As for South America, which has yielded most of the known remains of Phorusrhacidae, the studies about their paleoenvironments have focused principally on the Neogene formations of Argentina. Degrange *et al.* (2012) studied the avifauna of the Santa Cruz Formation (the upper part of the Lower Miocene), which has yielded a particularly diverse assemblage of Phorusrhacidae (see above). They concluded that these birds inhabited an environment similar to the present-day humid Chaco, where open areas and more forested zones alternate along rivers and marshes. Phorusrhacidae, especially middle-sized to large species, are thought to have preferred open areas (in keeping with their cursorial type of locomotion). Tambussi (2011) emphasized the open nature of the environments suggested by the abundance of curosrial birds, while also acknowledging the presence of more forested areas.

As Tambussi (2011) pointed out, climate conditions significantly deteriorated in Patagonia from the Late Miocene onward, with lower temperatures and increased aridity (see also Tambussi and Degrange, 2013).

Therefore, Phorusrhacidae from the Huayquerian (Late Miocene) must have lived in environments characterized by open grasslands, where they coexisted with other cursorial birds such as Rheidae.

The climate trend that began at the end of the Miocene persisted during the Pliocene. Thus, in the Chapadmalalan, we find Phorusrhacidae together with a robust rheid, *Hinasuri*, in a steppe environment (Tambussi, 2011).

The gradual aridification of extra-Andine Patagonia during the Neogene does not seem to have had a harmful effect on Phorusrhacidae, which were well adapted to open environments, especially thanks to their locomotion type.

5.6. Extinction

We only have a very limited knowledge of African and European Phorusrhacidae, whose extinction date is uncertain. This is especially true for Africa, where the fossil record of vertebrates during the Eocene is still quite incomplete. In Europe, *Eleutherornis* is only known from the Lutetian (Middle Eocene). The rich avifauna from the Quercy Phosphorites (mainly from the Upper Eocene to the Upper Oligocene) does not include any Phorusrhacidae, and the presence of this group seems to have been short-lasting. The development of carnivorous mammals in Europe during the Eocene may have played a part in the disappearance of these birds.

The age of the last South American Phorusrhacidae remains a matter of discussion. Tambussi *et al.* (1999) described a tibiotarsus belonging to a large specimen of Phorusrhacidae from the Raigón Formation of Uruguay, which is referred to a period ranging from the Pliocene to the Early Pleistocene (a Pliocene age cannot be excluded according to Agnolin (2013)). More recently, Alvarenga *et al.* (2010) have attributed an incomplete tarsometatarsus from the Dolores Formation of Uruguay (Middle to Upper Pleistocene) to a fairly small individual of Phorusrhacidae. However, Tambussi (2011) and Agnolin (2013) contested it. Jones *et al.* (2016) have recently reported additional specimens (humerus, tarsometatarsus) attributed to Psilopterinae from the Dolores Formation, for which a date of 96040 ± 6300 years was obtained thanks to optically simulated luminescence. If these specimens have been identified correctly, then Phorusrhacidae may have lived on in Uruguay until the Late Pleistocene.

The phorusrhacid species known from North America, *Titanis walleri*, was described by Brodkorb (1963) on the basis of remains found in sediments initially attributed to the Upper Pleistocene in Florida. Baskin (1995) referred a phalanx from Texas to the Late Pleistocene (Rancholabrean). However, in their review of *Titanis walleri*, Gould and Quitmyer (2005) refered all the material found in Florida to the Pliocene (Blancan) and concluded that the Texan specimen is also Pliocene in age. On the basis of studies about the rare elements contained in fossils, MacFadden *et al.* (2007) concluded that *Titanis walleri* reached North America at the beginning of the Pliocene, even before a land connection was established between North America and South America, and that it became extinct in North America before the beginning of the Pleistocene. Therefore, the idea occasionally mentioned by the media according to which this giant bird might have co-existed with the first human beings who reached North America is completely groundless.

The causes behind the disappearance of American Phorusrhacidae remain unclear. The fact that these animals figure among the forms that spread from South America to North America during the Great American Faunal Interchange in the Pliocene should be taken into consideration, as it does not suggest that Phorusrhacidae fell immediately victim to the competition or predation of animals that came from the north. Nonetheless, their presence in North America seems to have been short-lived. According to MacFadden *et al.* (2007), *Titanis* may have disappeared due to the competition of large carnivorous mammals (canids, felids and ursids), which diversified in the region during the Pliocene. If the presence of small Phorusrhacidae in Uruguay during the Late Pleistocene is confirmed, large forms still present in South America at the end of the Pliocene and even at the beginning of the Pleistocene may have disappeared earlier than smaller forms, maybe falling victim to the competition of large placental carnivores, whereas small forms may have persisted longer. Tonni and Noriega (1998) suggested that the disappearance of Phorusrhacidae, at least on a local level, may have been caused by the competition of condors, but this explanation implies that Phorusrhacidae were mostly scavengers, which remains to be proven.

6

Brontornithidae

6.1. Historical background

Brontornis burmeisteri was described by Moreno and Mercerat (1891) on the basis of several parts of a skeleton, including several bones of the hind-limb of the same individual, found in the Santa Cruz Formation in the Argentinian part of Patagonia. Later, Ameghino (1895) referred additional specimens to this species. Moreno and Mercerat classified *Brontornis* in a distinct family, Brontornithidae, placed among Stereornithes, an order that essentially included birds now considered as Phorusrhacidae. Dolgopol de Saez (1927) highlighted different features that separated Brontornithidae from Phorusrhacidae, and placed them in a separate order, Brontornithes. This interpretation was accepted by Kraglievich (1932). Ever since, there have been two opposite approaches concerning *Brontornis*: some authors (Patterson & Kraglievich, 1960; Brodkorb, 1967; Acosta Hospitaleche *et al.*, 2001; Alvarenga & Höfling, 2003; Alvarenga *et al.*, 2011) regarded it as a particularly massive phorusrhacid, whereas others think that it represents a distinct group, with no close relation to Phorusrhacidae. In particular, Agnolin (2013, 2017) considers *Brontornis* as a giant basal anseriform. Tambussi (2011), Tambussi and Degrange (2013), and Buffetaut (2014), as well as the authors of this book, accept this interpretation. In any case, all the authors agree to consider *Brontornis* as the most massive of all the South American land birds.

6.2. Taxonomy, systematics and classification

6.2.1. *Taxonomy*

Two species of *Brontornis* have been described, *B. burmeisteri* Moreno and Mercerat, 1891 and *B. platyonyx* Ameghino, 1895. The material from the Santa Cruz Formation described by Ameghino (1895) under the latter name may actually be referred to *B. burmeisteri* (Alvarenga & Höfling, 2003). Apart from *Brontornis burmeisteri*, several other taxa, in most cases represented by fragmentary remains from the Santa Cruz Formation, have been placed among Brontornithidae, especially *Rostrornis floweri* Moreno & Mercerat, 1891 and *Liornis floweri* Ameghino, 1895 (Dolgopol de Saez, 1927). Both *Rostrornis floweri* (Alvarenga & Höfling, 2003) and *Liornis floweri* (Brodkorb, 1967; Buffetaut, 2016b) can be considered as junior synonyms of *Brontornis burmeisteri*. *Callornis giganteus* Ameghino, 1895 is a chimera based on a tibiotarsus belonging to a brontornithid and a tarsometatarsus belonging to a phorusrhacid (Buffetaut, 2016b). *Liornis minor*, which Dolgopol de Saez (1927) considered as a gracile brontornithid on the basis of an isolated tarsometatarsus, is apparently a phorusrhacid which may be referred to *Phorusrhacos longissimus* (Alvarenga & Höfling, 2003; Buffetaut, 2016b). Therefore, it seems that all the material belonging to Brontornithidae from the Santa Cruz Formation may be referred to *Brontornis burmeisteri* or regarded as Brontornithidae *indet.* when the material available is too fragmentary to be identified with precision. The specimens attributable to Brontornithidae from other formations (especially those from the Oligocene of Bolivia) are too incomplete to be identified beyond Brontornithidae *indet.*

6.2.2. *Systematics*

Class Aves Linnaeus, 1758

Family Brontornithidae Moreno & Mercerat, 1891

Genus *Brontornis* Moreno & Mercerat, 1891

– *Brontornis burmeisteri* Moreno & Mercerat, 1891

Stratigraphic occurrence: Lower Miocene (Santacrucian) of Argentina (Patagonia, Santa Cruz Formation) and Upper Oligocene (Deseadan) of Bolivia (Salla Beds).

6.2.3. *Classification*

We regard Brontornithidae as a family distinct from Phorusrhacidae, which can be placed, as suggested by Agnolin (2013, 2007), among basal Anseriformes. In keeping with the phylogeny put forward by Agnolin (2007), *Brontornis burmeisteri* would be closely related to Gastornithidae (from Laurasia) and Dromornithidae (from Australia) (Figure 6.1).

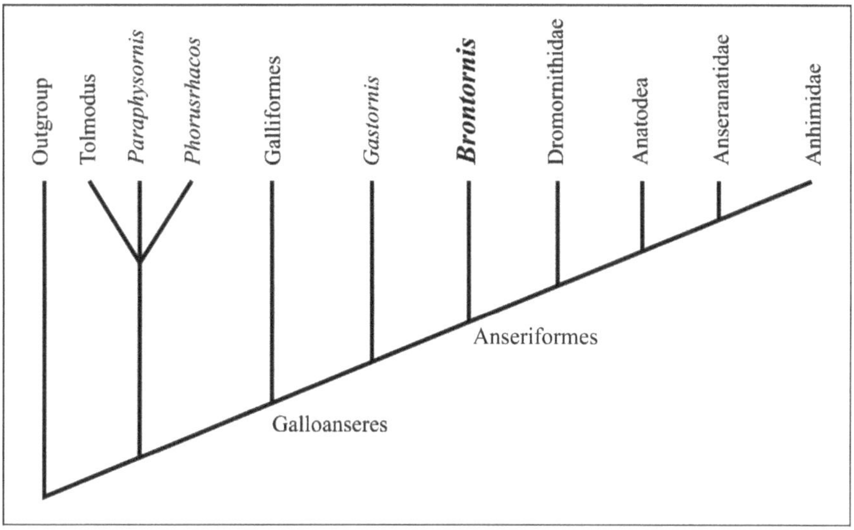

Figure 6.1. *Phylogenetic position of* Brontornis, *according to Agnolin (2007)*

6.3. Stratigraphic and geographical distribution

The vast majority of the remains of Brontornithidae identified so far come from the Santa Cruz Formation in the Argentinian part of Patagonia (provinces of Santa Cruz and Chubut), which corresponds to the upper part of the Lower Miocene (South American Land Mammal Age – SALMA – Santacrucian, 19 to 14 Ma; Perkins *et al.*, 2012) (Figure 6.2). From the "*Colpodon* beds" of Patagonia (Colhuehuapian, middle part of the Lower Miocene), Ameghino (1901) reported the presence of *Brontornis modestus*, which must be regarded as a *nomen dubium*, as no description or illustration has been provided and the present location of the material is unknown.

The presence of a brontornithid in the Salla Beds (Upper Oligocene, Deseadan SALMA, approximately 25–26 Ma; Kay *et al.*, 1998) in the

Salla-Luribay basin in Bolivia is indicated by a fragment of tibiotarsus which clearly exhibits the features characteristic of this family (Buffetaut, in press).

Figure 6.2. *Geographical distribution of the main fossil sites where specimens of Brontornithidae have been found. Lower Miocene site: 1 – Santa Cruz Formation (Argentina) and Upper Oligocene site: 2 – Salla-Luribay Beds (Bolivia)*

Therefore, Brontornithidae are currently known only from the Late Oligocene and the end of the Early Miocene in Argentina and Bolivia (Figure 6.3). It can apparently be regarded as a family endemic to South America; however, its evolutionary history is mostly unknown due to a very limited fossil record.

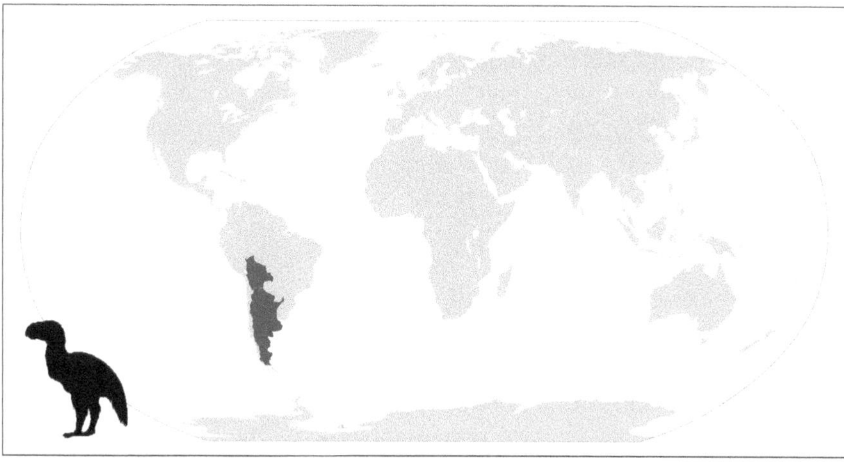

Figure 6.3. *Geographical distribution of the countries where brontornithid remains have been found. For a color version of the figure, see www.iste.co.uk/angst/birds.zip*

6.4. Anatomy

Figure 6.4. *Skeleton and restoration of* Brontornis

Our knowledge of the osteology of Brontornithidae is incomplete, and several parts of their skeleton remain unknown (Figure 6.4). In particular, we know virtually nothing about their skull. The quadrate, described in detail by Agnolin (2013, 2007), reveals only two condyles, as the caudal condyle is reduced (Figure 6.5(A)). This is a synapomorphy of Galloanseres that separates *Brontornis* from Phorusrhacidae.

The mandible of *Brontornis burmeisteri* is represented by several specimens, none of which is complete. The best specimen, kept at the National Museum of Natural History in Paris (Figure 6.5(B)), shows a relatively short and high symphysis, which clearly differs from the low and elongate symphysis of Phorusrhacidae. The mandibular rami are incompletely preserved but are remarkably high. The external mandibular fenestra is either absent or very reduced, unlike that of Phorusrhacidae, which is large and elongated. The attachment surface for the adductor muscles on the mandible seems to have been large, which is not the case in Phorusrhacidae.

The thoracic vertebrae attributed to *Brontornis* exhibit pneumatic cavities on the lateral surfaces of the centrum, which Alvarenga *et al.* (2011) regarded as a synapomorphy of Phorusrhacidae. However, Agnolin (2013) pointed out that it represents a plesiomorphic feature in Neornithes, present in numerous Galliformes and Anseriformes.

The hindlimb of *Brontornis burmeisteri* is fairly well known. The femur is short and very robust (Figure 6.5(C)). The tibiotarsus is elongate and presents a very strong cnemial crest (Figure 6.5(E)). There is no ossified supratendinal bridge, which is a feature that separates Brontornithidae from Phorusrhacidae; instead, a system of tubercles and ridges is present. The tarsometatarsus is short, wide and craniocaudally flattened (Figure 6.5(D)). Unlike Phorusrhacidae, the *canalis interosseus distalis* is not bifurcated and has no opening on the caudal surface of the bone. The phalanges are short and robust (Figure 6.5(F)). The ungual phalanges, which differ from the claws of Phorusrhacidae, have a posterior articular surface shaped like a wide-based triangle and are not markedly hooked (Figure 6.5(E)).

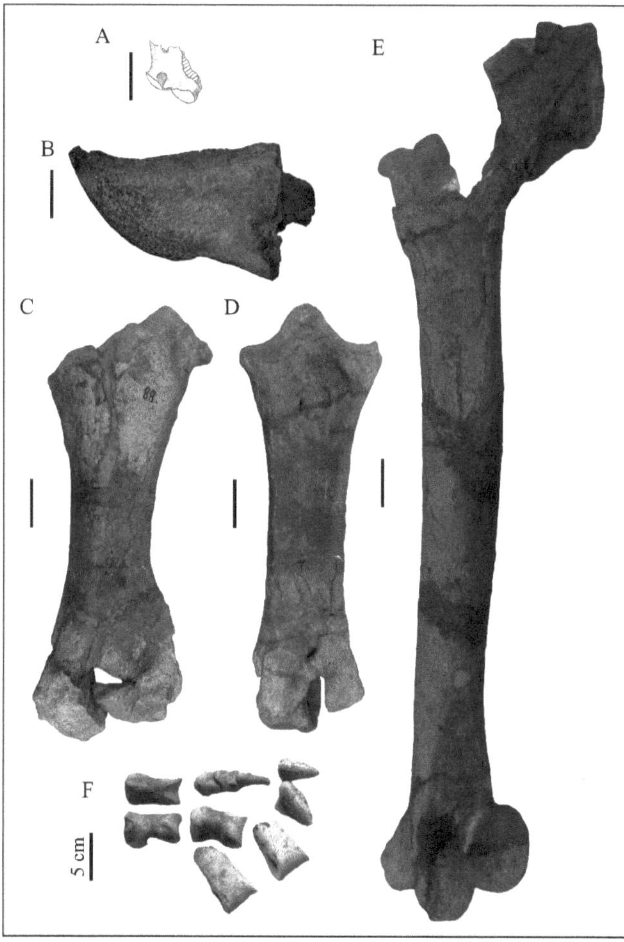

Figure 6.5. *Pictures of the main known bones of* Brontornis. *(A) Quadrate (after Agnolin (2007)), (B) Mandible (according to Agnolin (2007) and Buffetaut (2014), (C) Femur (Photos by Angst and Buffetaut), (D) Tarsometatarsus (Photos by Angst and Buffetaut), (E) Tibiotarsus (Photos by Angst and Buffetaut) and (F) Phalanges (according to Alvarenga & Höfling (2003)). For a color version of the figure, see www.iste.co.uk/angst/birds.zip*

6.5. Paleophysiology and paleoecology

6.5.1. *Diet*

Hypotheses about the lifestyle of *Brontornis* have been influenced by the idea that it merely represents an especially massive phorusrhacid. As Phorusrhacidae, in general, were carnivorous, this type of diet has been frequently attributed to *Brontornis*. Trouessart (1895) also suggested that it may have fed on dinosaurs, thus overestimating both its geological age and dimensions! More recently, Alvarenga & Höfling (2003) suggested that Brontornithidae (among which they also place *Paraphysornis*, which is clearly a phorusrhacid), were scavengers, in line with their heavy build and their type of locomotion, which was probably slow. In keeping with a carnivorous diet, some restorations (e.g. Alvarenga *et al.*, 2011) showed this bird with a hooked beak similar to that of Phorusrhacidae. However, the upper jaw of *Brontornis* is unknown and there is no evidence that it ended in a hook similar to that of birds of prey. The lower jaw, which is deep and wide, is reminiscent of Gastornithidae and Dromornithidae, which are now considered as herbivorous birds, and some recent restorations (Degrange *et al.*, 2012; Tambussi, 2011) attribute to *Brontornis* a beak similar to that of these birds, even though these authors hesitate to make claims about their diet. As mentioned above, the traces of muscular attachment on the mandible seem to indicate a noticeable development of the adductors (Buffetaut, 2014). As Angst *et al.* (2014) demonstrated, these muscles are strong in herbivorous birds, but relatively underdeveloped in carnivorous forms. What is known about the lower jaw of *Brontornis* therefore suggests that its diet was herbivorous, as already suggested by Agnolin (2007). Similarly, the broad and weakly curved phalanges differ from the claws of Phorusrhacidae and other carnivorous birds.

6.5.2. *Body mass*

Alvarenga and Höfling (2003) estimated that *Brontornis burmeisteri* reached a height of approximately 175 cm measured at the level of its back and approximately 280 cm measured at the level of its head with the neck raised. The mass estimated by these authors ranges from 350 to 400 kg (based on comparisons with present-day birds of known size and body mass), which makes *Brontornis* one of the largest birds that have ever lived, together with *Dromornis stirtoni* and *Aepyornis maximus*. The estimates made by Degrange *et al.* (2012) on the basis of the circumference of the femur and tibiotarsus (Campbell and Marcus method (1992)) yield a value of 319 kg and 350 kg, respectively.

6.5.3. *Locomotion*

As to the type of locomotion, the relatively short and wide tarsometatarsus of *Brontornis* is significantly different from the long and slender one of most Phorusrhacidae. Following the method put forward by Angst *et al.* (2016) to estimate the type of locomotion of giant birds based on the proportions of their tarsometatarsus, *Brontornis burmeisteri* can clearly be classed among the graviportal forms, which are incapable of rapid running (ratio equal to 5.5; Figure 6.6). This represents an additional argument in favor of the herbivorous diet of Brontornithidae, which seem to have been unable to chase rapidly running prey, unlike most Phorusrhacidae, which were evidently cursorial. Nevertheless, it should be pointed out that the phorusrhacids *Paraphysornis* and *Kelenken*, whose skeleton is very robust, are also classed among the graviportal forms; however, their mandible does not essentially differ from that of other Phorusrhacidae (they were very likely specialized carnivores which did not chase their prey).

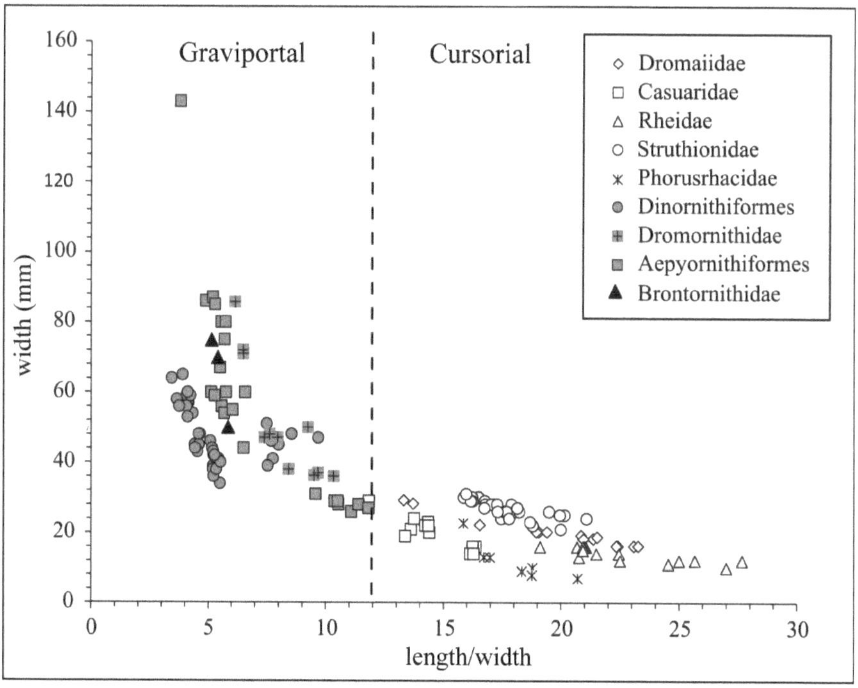

Figure 6.6. *Estimate of the type of locomotion of Brontornithidae based on the method used by Angst* et al. *(2016)*

6.5.4. *Sexual dimorphism*

Having observed significant differences in size between the tarsometatarsi of *Brontornis burmeisteri*, as some bones are up to 33% smaller than others, Alvarenga and Höfling (2003) suggested the existence of sexual dimorphism in this species. This hypothesis is interesting and deserves to be tested with more thorough statistical studies. Histological investigations could show, through the discovery of medullary bone, whether the largest individuals were males (as is the case for ostriches) or females (as was the case for some moa).

6.6. Paleoenvironment

These considerations result in an image of Brontornithidae as large graviportal birds with a herbivorous diet, quite different in this respect from Phorusrhacidae, with which they shared their environment. The ecological role of Brontornithidae during the Tertiary in South America seems quite similar to that of Gastornithidae in the Early Eocene of the Northern Hemisphere and of Dromornithidae in the Cenozoic of Australia, namely that of large herbivores. In the Deseadan of Bolivia, as well as in the Santacrucian of Patagonia, they were part of a fauna that included various herbivorous mammals, among which some large-sized ones. Land carnivores were mainly marsupials and birds (Phorusrhacidae). During the Santacrucian, Patagonia seems to have hosted a mixture of open and closed plant environments, in a climate that exhibited a certain degree of seasonality (Tambussi, 2011; Degrange *et al.*, 2012). Degrange *et al.* (2012) suggested that the paleoenvironment resembled that of the present-day humid Chaco and thought that middle-to-large Phorusrhacidae preferred open habitats due to their cursorial type of locomotion. Brontornithidae, whose locomotion was graviportal, may have preferred a more closed type of habitat, which could provide them with more abundant plant food.

6.7. Extinction

The most recent Brontornithidae known are those found in the Santa Cruz Formation (end of the Upper Miocene). The remains of very large land birds have been identified in several more recent South American formations, but they belong to Phorusrhacidae, and currently there is no fossil evidence suggesting the existence of Brontornithidae after the Early Miocene. The causes behind their disappearance are unclear. It may be assumed that it was associated with the climate changes that took place in South America after the Early Miocene, characterized by a trend towards a cooler climate and increasing aridity (Tambussi & Degrange, 2013), which caused a decrease in forested areas to the advantage of steppes. If Brontornithidae were adapted to relatively closed environments, this evolution may have caused their decline. Nevertheless, it should be pointed out that this evolution of the environment was more marked in the southernmost regions of South America and that what may possibly account for the disappearance of Brontornithidae in Patagonia does not necessarily apply to zones farther north.

7

Gastornithidae

7.1. Historical background

The first fossil attributed to the Gastornithidae was a tibiotarsus discovered in 1855 by Gaston Planté in the Meudon Conglomerate near Paris (Hébert, 1855a; Lartet, 1855; Prévost, 1855a), which led to the erection of the genus *Gastornis* in honor of its discoverer. More than 20 years later, additional specimens were discovered in the Upper Paleocene of France by Victor Lemoine (1878, 1881), and then in other parts of Europe. At the same time in North America, Cope (1876) described a new fossil of a giant bird as *Diatryma* while indicating that it was related to *Gastornis*. More recently, Hou (1980) has described the first and only Chinese specimen referred to this family. For several decades, all the studies carried out on the Gastornithidae focused on anatomical descriptions and issues of phylogeny or taxonomy, with only a few ecological considerations. The paleoecology of this bird has been widely studied since the 1980s and then again since 2013, which led to a better understanding of the ecology and biology of these large terrestrial birds, which lived at the beginning of the Tertiary, just after the extinction of non-avian dinosaurs.

7.2. Taxonomy, systematics and classification

7.2.1. *Taxonomy*

Four genera have been described within the family Gastornithidae: *Gastornis*, *Diatryma*, *Omorhamphus* and *Zhongyuanus*. The genus *Gastornisi* was the first one to be erected on the basis of a tibiotarsus discovered in 1855 in the Meudon Conglomerate, near Paris (Hébert, 1855a; Lartet, 1855; Prévost, 1855a). The genus was named *Gastornis* after its discoverer Gaston Planté, who was an assistant lecturer in physics at the Conservatoire des Arts et Métiers and an amateur paleontologist. Subsequently, more material referred to this genus was discovered in the Upper Paleocene of France by Victor Lemoine (1878, 1881), leading him to publish a restoration of *Gastornis* (Lemoine, 1878) (Figure 7.1(A)). At the same time, in North America, Cope (1876) discovered a new fossil of a giant bird that he regarded as related to *Gastornis*, but he noted enough anatomical differences to give it a new generic name, *Diatryma*. For several decades, both names were used, *Gastornis* being used in Europe (Dollo, 1883; Newton, 1885a, 1885b, 1886; Dollo, 1909) and *Diatryma* in North America (Cope, 1876, 1877). In 1917, Matthew and Granger (1917) described a sub-complete skeleton of *Diatryma* from the Lower Eocene of Wyoming, which made it possible to provide a global restoration of this bird (Figure 7.1(C)). The especially marked differences between this restoration and that of *Gastornis* published by Lemoine led to a clear-cut separation between the genera *Diatryma* and *Gastornis*. Furthermore, the excellent preservation of the American fossil largely eclipsed *Gastornis*. This is why most European specimens from the 1920s onward were attributed to *Diatryma* rather than *Gastornis* (Schaub, 1940; Fischer, 1962; Berg, 1965; Fischer, 1978), with the exception of the German specimens from Walbeck reported by Weigelt (1939). At the end of the 1920s, Sinclair erected a new genus within Gastornithidae, *Omorhamphus* (Sinclair, 1928a), which was synonymized with *Diatryma* by Andors several decades later (Andors, 1988, 1992). The coexistence of these three genera and the preferential attribution of gastornithid fossils to *Diatryma* rather than to the other two genera lasted until the beginning of the 1990s, when Larry Martin (1992) published a global revision of the *Gastornis* material used by Lemoine for his restoration. He highlighted then that the vast majority of the bones used did not belong to a bird (Figure 7.1(B)). The main argument that made it possible to distinguish between *Gastornis* and *Diatryma* was thus challenged. Martin suggested that these two genera should be grouped into the same family, Gastornithidae. The review of the fossil material attributed to these two genera shows that it belongs to a single genus (Buffetaut, 1997a; Angst & Buffetaut, 2013) which, following the priority rule of the International Code of Zoological Nomenclature, must be called *Gastornis* (*Diatryma* being considered as a junior synonym) as had already been suggested several times (Buffetaut, 1997a, 2000; Mlíkovský, 2002; Buffetaut, 2008; Mayr, 2009).

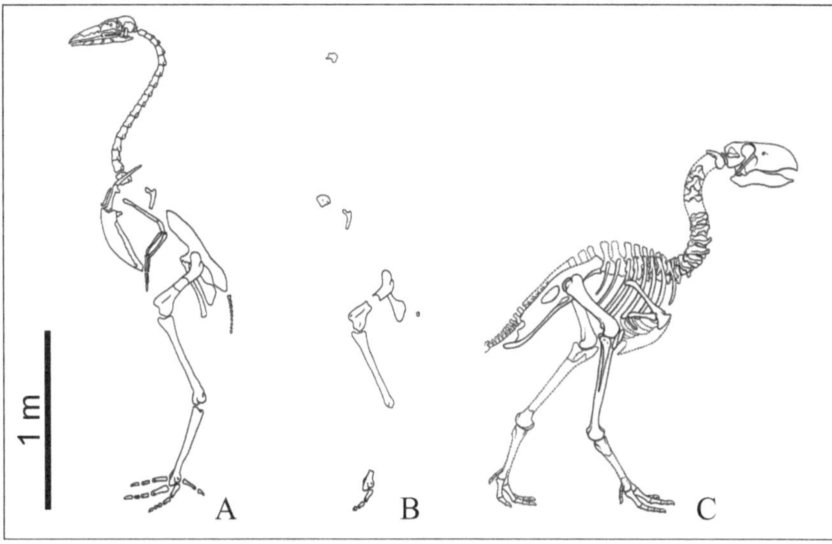

Figure 7.1. *Restoration of* Gastornis. *(A) Lemoine's restoration of* Gastornis *(1881), (B) Elements actually belonging to a bird in Lemoine's restoration (Martin, 1992), (C) Restoration of* Diatryma *(Matthew & Granger, 1917)*

More than 10 species have been described in the family Gastornithidae since 1885. All the North American species and part of the European ones were attributed to the genus *Diatryma*, while the rest of the European species were referred to the genus *Gastornis*. To clarify and simplify this particularly chaotic classification, several studies have been carried out in the last 20 years. All the North American fossils of *Diatryma* were studied again by Andors, who synonymized most of the species described and kept only two of them: *Diatryma giganteus* and *Diatryma regens* (Andors, 1988, 1992). Unfortunately, he did not compare the North American taxa with the European species of *Diatryma*. A second study, carried out by Martin (1992), reviewed the European species of *Gastornis* from the Paleocene of France. Martin grouped all the existing species into a single species, *Gastornis parisiensis*, erected a new species, *Gastornis russelli* and regarded the species *Gastornis minor* as a *nomen dubium*, as the material referred to it was not sufficiently diagnostic. Once again, however, he did not make any comparisons with the European species of *Diatryma*. Therefore, three species remained excluded from all the taxonomic reviews hitherto carried out: *Diatryma cotei*, *Diatryma geiselensis* and *Diatryma sarasini*.

A recent study (Angst *et al.*, 2013) demonstrated that the material attributed to *Diatryma cotei*, which is now combined with *Eleutherornis helveticus*, does not belong to Gastornithidae and actually belongs to Phorusrhacidae. Two recent studies (Mlíkovský, 2002; Hellmund, 2013) reconsidered the taxonomic position of *Diatryma sarasini* Schaub, 1929 described on the basis of a phalanx and the distal end of a tarsometatarsus, and *Diatryma geiselensis* Fischer, 1978, which included cranial (mandible, premaxilla) and postcranial material (vertebrae, femur, sacrum, tibiotarsus, tarsometatarsus and phalanx). Both studies confirmed the synonymy of *Diatryma* and *Gastornis*; on the contrary, they drew contradictory conclusions about the specific attribution of this taxon. Mlíkovský (2002) synonymized the species *Diatryma geiselensis* and *Gastornis sarasini*, arguing that the material referred to *G. sarasini* does not allow us to dissociate it from *D. geiselensis*, and that these two taxa have similar stratigraphic and geographical occurrences, as well as anatomical features. Therefore, he suggested a new combination that, in keeping with the priority principle, must be *Gastornis sarasini*, as the specimen referred to *G. sarasini* Schaub, 1929 was described well before *D. geiselensis* Fischer, 1978. A recent study of the material initially referred to *D. geiselensis*is challenges this conclusion (Hellmund, 2013). According to this author, the type material of *G. sarasini* is too limited (a phalanx and the distal end of a tarsometatarsus) to justify a synonymy with *G. geiselensis*. He suggested that the two species should be kept, but no new anatomical evidence was provided to back up the species initially described as *G. sarasini* and dissociate it from *G. geiselensis*. Finally, a fourth genus, *Zhongyuanus* Hou, 1980, which includes only one species, *Z. xichuanensis*, was described within Gastornithidae. This species, from China, is known only from the distal end of a tibiotarsus (Hou, 1980; Buffetaut, 2013b). This fossil presents sufficient anatomical features to be included among Gastornithidae but there are not enough elements to dissociate it from the other fossils of this family at the genus level. This is why, as in previous cases, the genus *Zhongyuanus* was synonymized with the genus *Gastornis* (Buffetaut, 2013b). Nevertheless, a sufficient number of elements allow us to separate it from the other species under the combination *Gastornis xichuanensis*.

7.2.2. *Systematics*

Different systematic arrangements for Gastornithidae have been put forward since the 19th Century (Brodkorb, 1963; Andors, 1988; Martin, 1992; Mlíkovský, 2002; Mayr, 2009). The one put forward here is a synthesis of the latest studies on this group.

Class Aves Linnaeus, 1758

Super-order Anserimorphae Sibley *et al.*, 1988

Order Gastornithiformes Stejneger, 1885

Family Gastornithidae Fürbringer, 1888

Genus Gastornis Hébert, 1855a

– *Gastornis parisiensis* Hébert, 1855a

Stratigraphic occurrence: Upper Paleocene – Lower Eocene of France, England and Belgium.

– *Gastornis russelli* Martin, 1992

Stratigraphic occurrence: Upper Paleocene of France.

– *Gastornis sarasini* (Schaub, 1929)

Stratigraphic occurrence: Lower Eocene of France.

– *Gastornis geiselensis* (Fischer, 1978)

Stratigraphic occurrence: Middle Eocene of Germany.

– *Gastornis giganteus* (Cope, 1876)

Stratigraphic occurrence: Lower Eocene of the United States and Canada.

– *Gastornis regens* (Marsh, 1894)

Stratigraphic occurrence: Lower Eocene of the United States.

– *Gastornis xichuanensis* (Hou, 1980)

Stratigraphic occurrence: Lower Eocene of China.

7.2.3. *Classification*

The systematic position of Gastornithidae, partly summed up by Dollo (1883), Andors (1992), and Buffetaut (1997a), has given rise to several hypotheses since the discovery of bones in 1855. In the very first paper describing *Gastornis*, Hébert (1855a) associated it with "lamellirostrate palmipeds", which correspond to part of present-day Anseriformes, by comparing it to a swan, a goose or a duck. Milne-Edwards (1867) later took up this hypothesis. This genus was also placed among waders by Edouard Lartet (1855). Duméril (1855) compared it to a stork, and Owen (1856a) associated it with Grallatores (wading birds). Valenciennes (1855) compared it to albatrosses and Bonaparte, to the Aepyornithidae (1856). Later, Gastornithidae were referred to Paleognathae (ratites and tinamous) (Cope, 1876), Psittaciformes (parrots) (Andrews, 1917), Ciconiiformes (storks, marabous) (Troxell, 1931), Gruiformes (cranes, coots) (Matthew & Granger, 1917), and Anseriformes (ducks, geese) (Shufeldt, 1909). All these suggestions were reviewed by Andors (1988, 1992), who considered the Gastornithidae as the sister group of extant Anseriformes within the super-order Anserimorphae. Another more recent study, carried out by Agnolin (2007), placed *Gastornis* within Anseriformes *sensu stricto* at the base of the diversification of this group rather than as a sister group. Another hypothesis that has recently been put forward placed *Gastornis* among the Phorusrhacidae (Bourdon & Cracraft, 2011). This had already been suggested at the end of the 19th Century, when the Phorusrhacidae were placed among Gruiformes (Gadow, 1893; Ameghino, 1895; Lydekker, 1895; Zittel, 1902). Andrews' work (1899) demonstrated that the Phorusrhacidae are not Gruiformes, and Matthew and Granger (1917) rejected the hypothesis of a kinship between Phorusrhacidae and Gastornithidae. Bourdon and Cracraft's hypothesis also challenged the latest anatomical studies carried out in the 1990s by Andors (1988, 1992). Nevertheless, there are many anatomical differences between these two groups. Most Phorusrhacidae were cursorial birds with very slender hindlimb bones, whereas the Gastornithidae moved more slowly and had more robust leg bones (Angst *et al.*, 2016). Their cranial anatomy is also very different, as Phorusrhacidae have a very sharp and specialized beak, allowing them to grab their prey, and a mandible characterized by a large mandibular fenestra, whereas Gastornithidae have a wider beak and a mandible with no mandibular fenestra, giving them more crushing power. These main anatomical differences are not interpreted correctly in Bourdon and Cracraft's work, as these authors have based most of their anatomical comparisons between

Gastornithidae and Phorusrhacidae on *Brontornis* as a representative of Phorusrhacidae. This genus is now placed outside Phorusrhacidae by many authors, as it presents too many differences with the other genera of this family, and placed among Anseriformes (Agnolin, 2007, 2009, 2013; Angst & Buffetaut, 2013; Tambussi & Degrange, 2013). Therefore, the Gastornithidae may be related to the Brontornithidae, but this does not mean that they are Phorusrhacidae, leading back to the initial hypothesis: the Gastornithidae belong to the order Anseriformes. The only issue that must still be solved is to decide whether they belong to Anseriformes *stricto sensu*, as suggested by Agnolin (2007), or should be regarded as the sister group of extant Anseriformes within the super-order Anserimorphae, in line with Andors' work (1992).

7.3. Stratigraphic and geographical distribution

7.3.1. *Body fossils of Gastornithidae*

Figure 7.2. *Geographical distribution of the countries where body fossils of Gastornithidae have been found. For a color version of the figure, see www.iste.co.uk/angst/birds.zip*

The fossil material that has been attributed to Gastornithidae since the 19th Century comes from sites ranging from the Middle Paleocene to the Middle Eocene in Europe, North America and Asia (Figure 7.2, Figure 7.3).

The Palaeocene sites are exclusively European (Figure 7.3). The oldest site where *Gastornis* material has been found so far is Walbeck of Selandian age (MP 1-5). The most recent site is the Geiseltal, of Middle Lutetian age (MP 13). The following list includes the sites where remains of Gastornithidae have been found, more details being provided for particularly relevant sites.

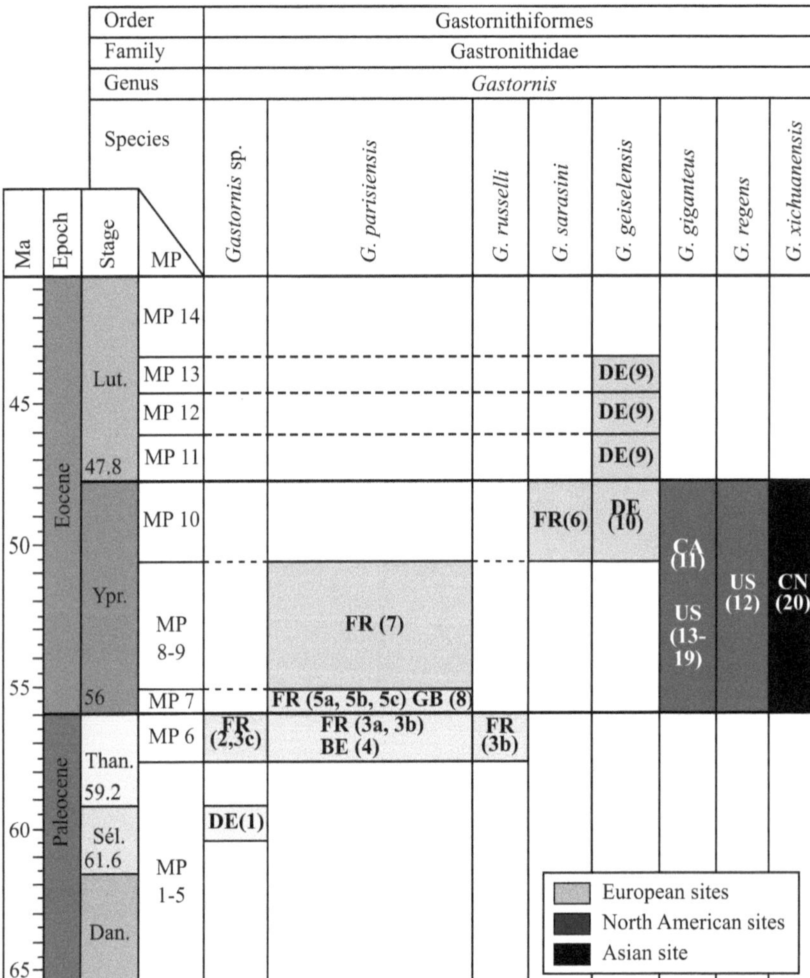

Figure 7.3. *Stratigraphic distribution of the body fossils of Gastornithidae. FR: France, DE: Germany, GB: Great Britain, BE: Belgium, CA: Canada, US: United States, CN: China, 1 – Walbeck, 2 – Rivecourt, 3a – Cernay-lès-Reims, 3b – Mont-de-Berru, 3c – Louvois, 4 – Mesvin, 5a – Meudon, 5b – Passy, 5c – Mutigny, 6 – Monthelon, 7 – Saint-Papoul, 8 – Croydon, 9 – Geiseltal, 10 – Messel, 11 – Ellesmere Island, 12 – New Jersey, 13–16 Wyoming, 17–18 Colorado, 19 – New Mexico, 20 – Liguanqiao*

7.3.1.1. *France*

7.3.1.1.1. Cernay-lès-Reims

This site is Late Paleocene in age (Thanetian, MP6; MP: Mammal Paleogene reference level) and has yielded remains of *G. parisiensis*, the first fossils referred to *Gastornis* having been described by Lemoine at the end of the 19th Century (1878, 1881).

7.3.1.1.2. Mont-de-Berru

The Mont-de-Berru site is Late Paleocene in age (Thanetian, MP6) and has yielded specimens of *G. parisiensis*. During the 1950s and the 1960s, excavations carried out on this site led to the discovery of several well-preserved fossils, described by Martin (1992), Buffetaut (1997a), and Angst & Buffetaut (2013). The Mont-de-Berru site also represents the only place where the two specimens described as *G. russelli*, a left tarsometatarsus and an upper jaw, were discovered.

7.3.1.1.3. Louvois

This site is Thanetian in age and corresponds to MP6, like the Cernay and Berru sites (Duprat *et al.*, 2010). Louis (1996) pointed out that this site yielded various broken bones belonging to large vertebrates, including abundant material referred to *Gastornis*. All the specimens discovered at Louvois have been recently described by Mourer-Chauviré and Bourdon (2015).

7.3.1.1.4. Petit Pâtis (Rivecourt)

This site, which is Upper Paleocene in age (Thanetian, MP6), has been reported only very recently with the publication of Smith *et al.* (2014). It has yielded only one fossil referred to *Gastornis*, namely a very large avian zygapophysis, which has not yet been described or figured.

7.3.1.1.5. Monthelon

The Monthelon site was described by Schaub (1929), who noted that it had yielded material belonging to Gastornithidae. He described the type series of the species *Diatryma sarasini*, now called *Gastornis sarasini*. These fossils come from the Sables à Unios et Térédines Formation, which is Ypresian in age (MP10 a and b) (Duprat *et al.*, 2010).

7.3.1.1.6. Meudon

The Meudon Conglomerate is the site where the first fossil (a tibiotarsus) referred to the family Gastornitihidae (*Gastornis parisiensis*) was discovered by Gaston Planté in 1855 (Prévost, 1855b, 1855a; Hébert, 1855a, 1855b; Lartet, 1855; Buffetaut, 1997b). This site is Ypresian in age, more precisely corresponding to MP7.

7.3.1.1.7. Passy

The Passy site is considered as coeval with the Meudon site, that is, Lower Ypresian (MP7), and it yielded fragmentary *Gastornis* material, figured by Milne-Edwards (1867) and mentioned by Dollo (1883).

7.3.1.1.8. Mutigny

This site is Early Ypresian (MP7) in age and has yielded a single fibula (MU 12734) attributed to *Gastornis parisiensis* but never studied or figured (Martin, 1992; Louis, 1996; Mourer-Chauviré & Bourdon, 2015).

7.3.1.1.9. Saint-Papoul

This site has been referred to the Early Eocene (Ypresian) and, more precisely, to MP 8-9, thanks to the significant fauna found there, which includes invertebrates (mollusks, crustaceans), fish, amphibians, turtles, squamates, crocodiles and various mammals. The fossils of *Gastornis parisiensis* found there have been described in several papers (Buffetaut, 2008; Laurent *et al.*, 2010; Bourdon *et al.*, 2014).

7.3.1.2. *Belgium*

7.3.1.2.1. Mesvin

Mesvin is the only Belgian site that has yielded *Gastornis* material. This site is referred to the Landenian (Dollo, 1883, 1909), a Belgian local stage corresponding to the Thanetian (MP6) on the international stratigraphic scale (De Geyter *et al.*, 2006).

7.3.1.3. *England*

7.3.1.3.1. Croydon

This site belongs to the Woolwich Formation, of Early Eocene age. Its age has been estimated with more precision thanks to the discovery of an abundant fauna and flora associated with the fossils of *Gastornis* (Hooker *et al.*, 2009). The study of these fossils has shown that this site corresponds to the end of the PETM (Palaeocene-Eocene Thermal Maximum). A more detailed study of the fossil mammals found at this site suggests that the Croydon fauna is stratigraphically equivalent to the Meudon fauna, Early Ypresian (MP7, Hooker, 2012). The *Gastornis* material found at Croydon includes, among others, several fragmentary tibiotarsi and a femur, described and figured by Newton (1885b, 1886). On the basis of this study, Newton suggested that this material should be considered as a new species of Gastornithidae, *Gastornis klaasseni*, which is now synonymized with *Gastornis parisiensis* (Martin, 1992).

7.3.1.4. *Germany*

7.3.1.4.1. Walbeck

The oldest site where *Gastornis* material has been found so far is Walbeck, although only a coracoid is still available. This fossil site is considered as terminal Selandian in age, which corresponds to MP5.

7.3.1.4.2. Messel

The Messel site is known for its very rich and particularly well-preserved fossil assemblage, which includes plants as well as vertebrate and invertebrate animals. This fossil site has been dated thanks to both radiometric methods, according to which it dates back to 47.8±0.2 Ma (Mertz & Renne, 2005), and the study of its rich mammal fauna, which indicated that the site corresponds to MP11 (Franzen, 2005), that is, the beginning of the Lutetian (Middle Eocene). According to the most recent study by Lenz *et al.* (2015), the age of the Messel site ranges between 48.27±0.22 and 48.11±0.22, which is older than previously estimated and places this site in the Ypresian, just before the Ypresian/Lutetian boundary, that is, MP10. Surprisingly, the remains of *Gastornis* found there merely include a natural cast of a femur, which was studied in detail by Berg (1965), slightly smaller than the femora of *Gastornis* from North America. It was referred to the species *Diatryma cf. steini* (Berg, 1965), which is now synonymized with *Gastornis giganteus*. However, recent studies place this fossil in the same species as those from the Geiseltal, *Gastornis geiselensis* (Mlíkovský, 2002; Hellmund, 2013).

7.3.1.4.3. The Geiseltal

The Geiseltal site yielded a large number of fossils of Gastornithidae. The first fossils of *Gastornis* (initially called *Diatryma*) were described by Fischer (1962), who later described additional material (Fischer, 1978) on the basis of which he erected a new species, *Diatryma geiselensis*. A recent study by Hellmund (2013) provides a review of not only this material but also the stratigraphic distribution of the different fossil levels of this site. The stratigraphic review by Hellmund (2013) showed that the levels containing *Gastornis* in the Geiseltal range from the Unterkohle, which corresponds to MP11, to the Mittelkohle, which corresponds to MP13, that is, all the Lower and Middle Lutetian. Consequently, this is the most recent European site where *Gastornis* has been found and it corresponds to the last occurrence of body fossils attributed to Gastornithidae worldwide.

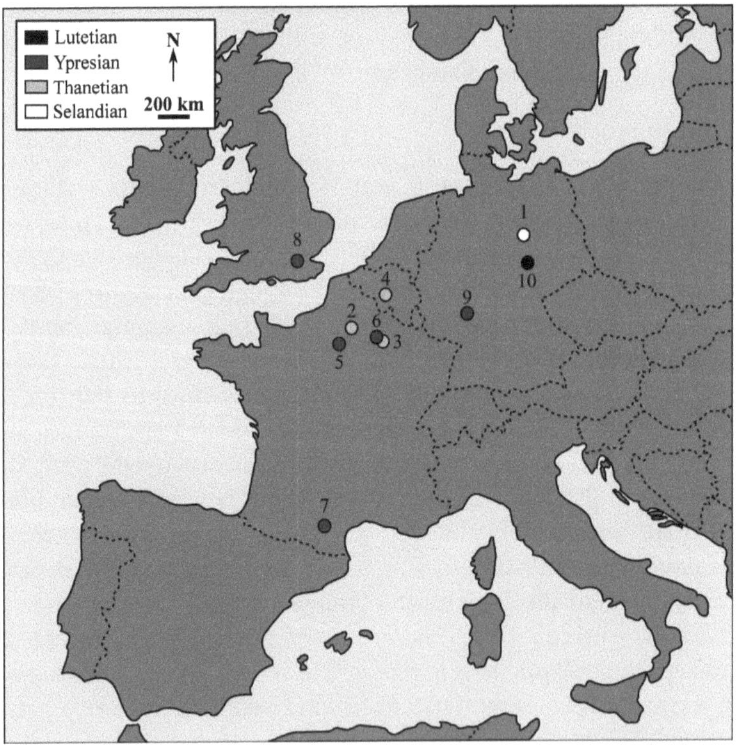

Figure 7.4. *Geographical distribution of the European sites. Selandian site: 1 – Walbeck; Thanetian sites: 2 – Rivecourt, 3 – Cernay-lès-Reims, Mont-de-Berru, Louvois , 4 – Mesvin; Ypresian sites: 5 – Meudon, Passy, 6 – Monthelon, Mutigny, 7 – Saint-Papoul, 8 – Croydon, 9 – Messel; Lutetian site: 10 – Geiseltal*

7.3.1.5. *USA*

7.3.1.5.1. New Mexico

New Mexico is the first state in the United States where material referred to Gastornithidae was found in 1874 during an expedition led by G. M. Wheeler (Andors, 1988). The first bone discovered was a tarsometatarsus, which was described by Cope (1876, 1877) and defined as the holotype of the type species of the genus *Diatryma* (species *D. giganteus*), now synonymized with the European genus *Gastornis*. The specimens come from the San Jose Formation in the San Juan basin in the north-eastern part of the state of New Mexico, dating back to the Early Eocene.

7.3.1.5.2. New Jersey

Marsh (1894) described a single phalanx from the Eocene upper marly layers, close to Squankum in Monmouth county (New Jersey) (Figure 7.5). On the basis of the remains of sedimentary matrix present on this fossil, Andors (1988) was able to establish that this fossil probably comes from the Squankum Member, of the Manasquan or Shark River Formation, both of them considered as Wasatchian in age (Early Eocene, Figure 7.3; Andors, 1992). Marsh erected a new genus and a new species, *Barornis regens*, for this fossil, which was re-attributed to the genus *Diatryma* by Andors (1988). Considering that two species could be distinguished within the genus *Diatryma*, Andors put forward the new combination *Diatryma regens*, which is now considered as a synonym of *Gastornis giganteus*.

7.3.1.5.3. Colorado

Two sites in the state of Colorado have yielded specimens of Gastornithidae, viz the DeBeque Formation in the Uinta Basin, in north-western Colorado, and the Huerfano A Formation, in Huerfano Park in the south-eastern part of the state. These two fossil sites are Wasatchian (Early Eocene) in age and have yielded relatively limited material, which was studied only briefly by Andors (1988), referred to *D. giganteus*, and never figured.

7.3.1.5.4. Wyoming

Wyoming is the state where the largest number of fossils have been found since the beginning of the 20th Century. The fossils come from four sites belonging to three geological formations.

Willwood Formation: most gastornithid specimens come from the Willwood Formation, in the north-western part of the state (Figure 7.5), and belong to the Wasatchian, which corresponds to the Ypresian (Rose, 1981) (Figure 7.3). This formation crops out in two sedimentary basins, the Bighorn Basin and the Clark's Fork Basin, with a larger proportion of the material coming from the former basin. A very large number of specimens have been discovered in these beds, including cranial as well as postcranial material, which was studied and described in detail by several authors since the beginning of the 20th Century (Matthew & Granger, 1917; Sinclair, 1928a, 1928b; Andors, 1988, 1992). Among these various specimens, some deserve to be examined in more detail. The material found in 1912, which contained phalanges and a trochlea, corresponds to the syntypes of the species *Diatryma ajax* described by Shufeldt (1913) and now synonymized with *Gastornis giganteus*. Furthermore, the type material of the species *Omorhamphus storchii* described by Sinclair (1928a, 1928b), which actually corresponds to a juvenile individual of *Gastornis giganteus* (Andors, 1988, 1992), includes several cranial and postcranial elements deriving from the Bighorn Basin. Finally, this formation has yielded the most complete skeleton of Gastornithidae hitherto discovered (n° AMNH. 6169), described and illustrated in detail by Matthew & Granger (1917) as *Diatryma steini* and eventually synonymized with *Gastornis giganteus*.

Wasatch Formation: the second geological formation in Wyoming that has yielded material referred to Gastornithidae is the Wasatch Formation, which crops out in the south-western part of the state and dates back to the Wasatchian (Ypresian). This formation crops out in two sedimentary basins, the Green River Basin and the Washakie Basin (Figure 7.5). The material from these two basins is more limited than that from the Willwood Formation (Andors, 1988).

Wind River Formation: the last geological formation is the Wind River Formation in the Laramie Basin, in the south east of the state (Figure 7.5). Only a few phalanges have been discovered at this site and then referred to *Diatryma giganteus* (Princhiello, 1971). This formation is Wasatchian in age (Ypresian) (Figure 7.3).

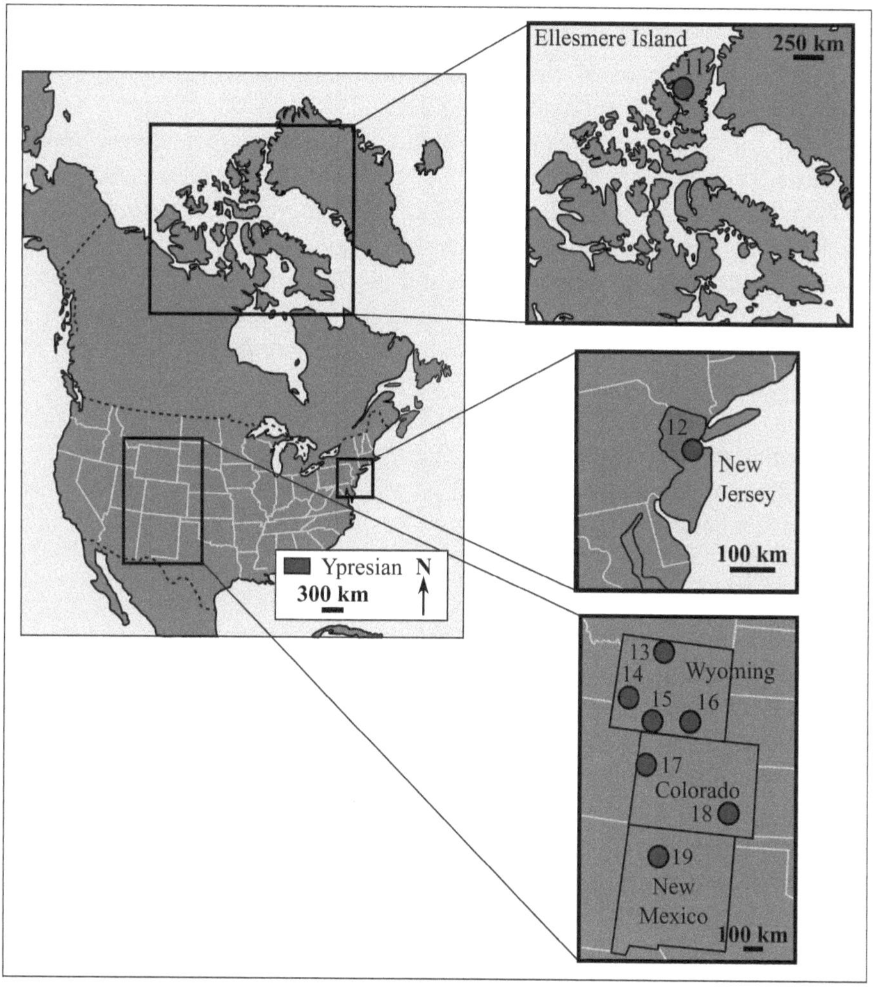

Figure 7.5. *Geographical distribution of the Eocene North American sites. Canada: 11 – Ellesmere Island; USA: 12 – New Jersey, 13-16 – Wyoming (13: Willwood Formation, 14: Wasatch Formation, Green River Basin, 15: Wasatch Formation, Washakie Basin, 16: Wind River Formation), 17-18 – Colorado (17: DeBeque Formation, 18: Huerfano A Formation), 19 – New Mexico*

7.3.1.6. *Canada*

The Canadian material comes from the central part of Ellesmere Island (Figure 7.5) and was referred by Andors (1988) to the species *Diatryma giganteus*, now renamed *Gastornis giganteus*. This specimen comes from the Sverdrup basin, of Late Wasatchian, i.e. Ypresian, age (Andors, 1988, 1992; Stidham & Eberle, 2016; Figure 7.3). The geographical position of this specimen is particularly interesting, as this fossil was found at a high latitude in a region halfway between Europe and North America and from a period when *Gastornis* is known to have lived in both regions. Moreover, the mammal fauna associated with this specimen presents similarities with Europe and North America, suggesting faunal dispersal between these two continents at the beginning of the Eocene. Therefore, the dispersal of Gastornithidae at the Paleocene-Eocene transition very probably took place via present-day northern Canada (West and Dawson, 1978; Andors, 1988; Stidham & Eberle, 2016).

7.3.1.7. *China*

The first description of material referred to Gastornithidae in Asia was published at the beginning of the 1980s (Hou, 1980). The fossil, the distal end of a tibiotarsus (V 5864), corresponds to the holotype of the species *Zhongyuanus xichuanensis*. It comes from the Liguanqiao basin, in the eastern part of the Qinling Mountains, in Henan in central China (Figure 7.6) (Buffetaut, 2013b). It was found in the Yuhuangding series, which is Early Eocene in age and has yielded various fossil mammals. Recent stratigraphic research places this site at the beginning of the Lower Eocene, at the level of the Bumbanian ALMA (Asian Land Mammal Age), which corresponds to the Wasatchian NALMA (North American land mammal), possibly corresponding to MP7 to MP9 at the beginning of the Ypresian (Vandenberghe *et al.*, 2012; Figure 7.3). The recent comparative study carried out by Buffetaut (2013b) places this species in the genus *Gastornis* as the new combination *Gastornis xichuanensis*.

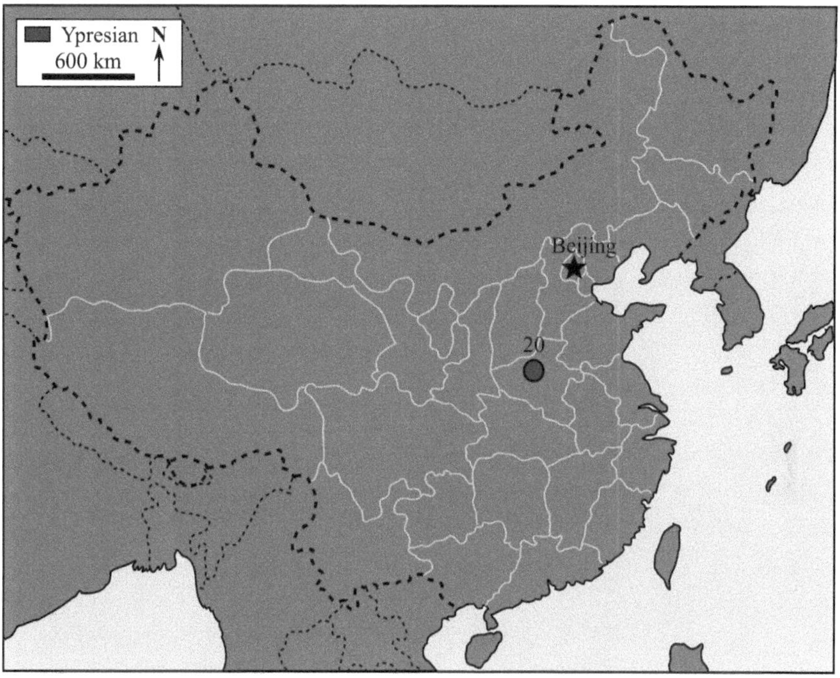

Figure 7.6. *Geographical location of the Early Eocene Chinese site. 20: Liguanqiao*

7.3.2. *Other fossils attributed to Gastornithidae*

Apart from the aforementioned body fossils, other types of fossil remains attributed to Gastornithidae have been found at other sites and geological periods.

Since the mid-20th Century, fragments of eggshells have been identified in the Tertiary sediments of southern France and northern Spain (Dughi & Sirugue, 1959; Touraine, 1960; Fabre-Taxy & Touraine, 1960; Corroy & Touraine, 1961; Touraine, 1961; Martini, 1961; Dughi & Sirugue, 1962; Touraine, 1962, 1963, 1967; Dughi & Sirugue, 1968; Touraine, 1978; Penner, 1983; Kerourio & Aujard, 1987; Donaire & Lopez-Martinez, 2009). Their geographical and stratigraphic occurrences are discussed in detail by Buffetaut and Angst (2014) (Figure 7.7). At these sites, two types of eggshells have been described: *Ornitholithus biroi* and *Ornitholithus arcuatus*. By studying the curvature of these eggshells, it has been possible to attribute only the latter (*O. arcuatus*), which are also the thicker ones, to Gastornithidae (Angst *et al.*, 2014b). These eggshells are known from the Eocene (Sparnacian) of 20 sites in southern France, in 11 sedimentary basins. They can also be found at three sites in northern Spain (Figure 7.7) (Buffetaut & Angst, 2014).

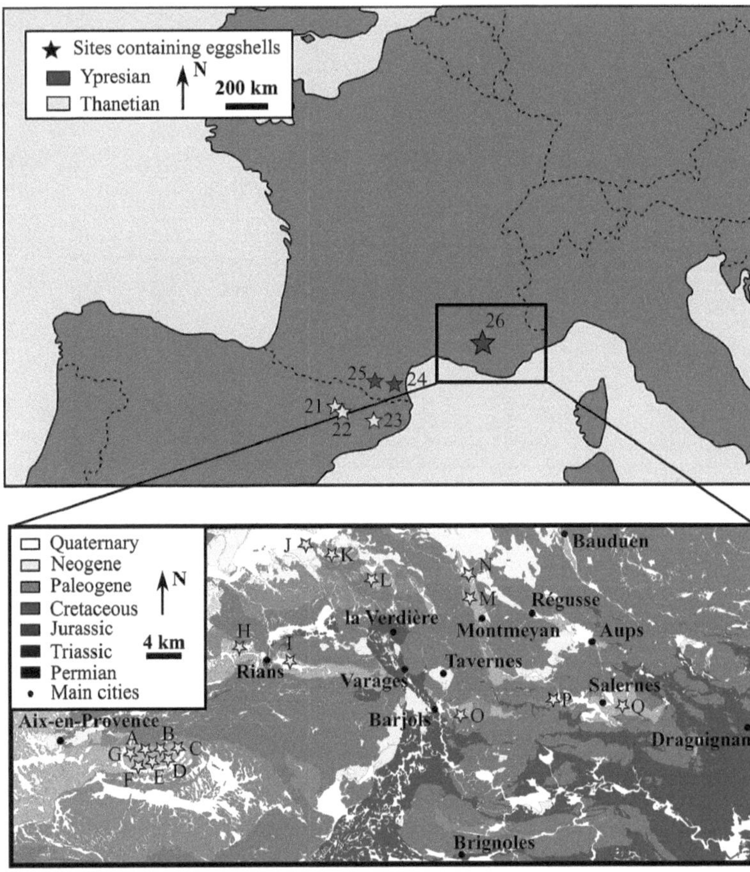

Figure 7.7. *Map of Europe showing the geographical position of the Paleogene sites that have yielded fossil eggshells of the* O. arcuatus *type. 21: Tremp, 22: Llimiana, 23: Manrèse, 24: Lagrasse, 25: Puivert, 26: Var-Bouches-du-Rhône. A – Saint-Antonin-sur-Bayon, B – Coquille, C – Suberoque, D – L'Etang, E – Bayle, F – Les Masques, G – Maurély, H – Les Bardouines, I – La Neuve, J – Vinon-sur-Verdon, K – La Baraque, L – Saint-Julien-le-Montagné, M – La Roquette, N – Sainte-Maime, O – Pontevès, P – Sillans-la-cascade, Q – Les Mauquiers*

The second type of fossil remains attributed to Gastornithidae are footprints. In 1859, French geologist Jules Desnoyers discovered footprints in France in the Late Eocene gypsum near Paris (Montmorency valley) (Desnoyers, 1859a, 1859b; Buffetaut, 2004). Desnoyers briefly described some bird footprints as large, consisting of three long and articulated fingers that ended with very pointed toes. In 1859, he suggested that they may be referred to *Gastornis*, noting, however, the stratigraphic problem that arose, as the most recent fossils of *Gastornis* were from the Lower Eocene (in Desnoyers' time, only the Meudon specimens were known) and these footprints were from the Upper Eocene, when *Gastornis* should have been extinct. Furthermore, Desnoyers noted the presence of two types of footprints of different sizes, revealing thus the presence of two birds. These footprints were later studied in detail by Milne-Edwards (1863, 1867, 1872), who confirmed the presence of two types of footprints of different sizes. Unfortunately, these footprints have never been figured and cannot be found at the moment and thus it is impossible to study them again (Buffetaut, 2004). Thanks to the North American material referred to *Gastornis*, which is particularly well preserved, it can be shown that a footprint left by this bird must have measured at least 20 cm in length (Andors, 1988). Thus, the particularly well-preserved footprints described by Desnoyers might have been left by *Gastornis*, as the other groups of land birds known to have lived in Europe in this period were too small to have left them. Nevertheless, European Gastornithidae were smaller than those from North America. Their footprints must have been correspondingly smaller, but it is difficult to provide an exact size. Besides, because of the lack of more precise descriptions of the French footprints, it is difficult to draw safe conclusions about their taxonomic attribution. Furthermore, the description of sharp nails at the end of the toes makes difficult an attribution to *Gastornis*. The terminal phalanges of *Gastornis* are not claws and more resembled small hoofs, which would then have left different footprints from those described by Desnoyers. Two hypotheses can then be put forward to account for this difference:

– The description of these footprints is not exact and the presence of claws is not proven. In that case, these footprints may have actually been left by *Gastornis*. Unfortunately, this cannot be verified, as the material has been lost. If we admit that this hypothesis is correct, it would push forward the date of the extinction of this group by several million years (up to the Late Eocene).

– The description is faithful to the fossil, so that these footprints cannot be attributed to *Gastornis*. This would tend to show that in France, during the Late Eocene, there were large birds with claws strong enough to leave their mark on the footprints. Among the birds known to have lived in Europe in this period and characterized by well-developed claws, only Phorusrhacidae are currently known. Although this group is so far represented in Europe only by a single and relatively small species, these footprints may demonstrate the presence of another taxon of larger Phorusrhacidae in Europe during the Late Eocene.

Considering the experience of the anatomists who have studied these footprints, the second hypothesis seems more plausible, even if it cannot be verified due to the lack of specimens.

In the United States, a footprint attributed to *Diatryma* (*Gastornis*) has also been described and referred to the Eocene. It was found in the Franklin Series of the Puget Group in the Green River Gorge, in the western part of Washington state, in laminated siltstone beds (Patterson & Lockley, 2004; Mayr, 2009). This single tridactyl footprint, which is 32.5 cm long, was found in 1992 by Patterson but actually studied only later due to a controversy about its authenticity (Patterson & Lockley, 2004). After the new study carried out by Patterson and Lockley (2004), this trace was attributed to a new ichnogenus, *Ornithoformipes controversus* and referred to *Diatryma* (*Gastornis*) because of a type of morphology that is completely in line with our knowledge of the foot of *Gastornis*. Nevertheless, the authenticity of this footprint was challenged by Mustoe *et al.* (2012), who suggested that it may actually represent a pseudo-fossil resulting from a possible geological transformation of the rock, instead of having been left by a living organism. If we assume that it represents an actual footprint, it is interesting to note that this footprint is estimated to be about 45 million years old (±2 Ma), corresponding to the Middle Eocene (Lutetian). As in Europe, this footprint is more recent than the latest bone remains attributed to *Gastornis* in North America, which date back to the Lower Eocene (Ypresian). No other bird large enough is known to have lived during the Eocene in North America, despite our very good knowledge of the Tertiary fossil record in the United States, raising the issue of the absence of *Gastornis* in these otherwise well-known formations. More recently, Mustoe *et al.* (2012) have described additional footprints from the Lower Eocene Chuckanut Formation in the western part of Whatcom County, in north-western Washington state. They attributed these well-preserved, tridactyl footprints to a new ichnogenus and species, *Rivavipes giganteus*, and referred it to *Gastornis*.

Finally, the last type of fossil that may be attributed to Gastornithidae is a 24 cm long feather discovered in the Green River Formation of Wyoming, in the United States, of Early Eocene age (Grande, 2013). Considering the stratigraphic and geographical distribution of the bone remains of Gastornithidae known from the United States, this large feather may be referred to this group.

7.4. Anatomy

Gastornithidae were large, massive terrestrial birds. The largest individuals (which are about 2 m tall) come from North America, whereas the European species were smaller (about 1.7 m tall) (Figure 7.8).

Figure 7.8. *Skeleton and restoration of* Gastornis

There is a very large number of gastornithid fossils, among which is a sub-complete skeleton of *G. giganteus*, kept at the American Museum of Natural History in New York, which was studied in detail by Matthew and Granger (1917) and more recently by Andors (1988). Not every specimen of Gastornithidae will be described in this section, which will only provide a general overview of the anatomy of Gastornithidae.

The back of the skull of these birds is known only thanks to the New York specimen. Even though the posterior part of the skull of this fossil is badly deformed, it can be seen that the skull is robust (Matthew & Granger, 1917; Andors, 1988) (Figure 7.9(A)). Several beaks belonging to Gastornithidae have been identified, among which are a specimen from the Geiseltal (Germany) (Fischer, 1978), another from Saint-Papoul (France) (Bourdon *et al.*, 2014), and yet another from Wyoming (USA) (Matthew & Granger, 1917; Andors, 1988), which is particularly well preserved. The beak is high and laterally compressed, and it does not present any hook at its tip. Seen laterally, the external nostril is small and in a dorsal and posterior position (Figure 7.9(B)). There are several more or less complete mandibles. One of the two best-preserved mandibles comes from Wyoming and is kept at the American Museum of Natural History in New York, whereas the other comes from Mont-de-Berru, and is kept at the Toulouse Natural History Museum. These two mandibles are interesting as they present complementary anatomical features linked to their preservation. The mandible presents a long tubular symphysis with high posterior rami and no external mandibular fenestra (Figure 7.9(C)).

The postcranial skeleton is represented by the vast majority of the known specimens. It includes massive but relatively short cervical vertebrae, which form a robust and relatively short neck (Figure 7.9(D)), providing information about the potential posture of this bird. The best-preserved pelvis referred to Gastornithidae belongs to the specimen kept in New York and is wide and robust (Figure 7.9(G)). The front limbs are very reduced and clearly did not allow these birds to fly. On the contrary, the hindlimbs are robust, with a short and wide tarsometatarsus (Figure 7.9(E)). Finally, the terminal phalanges are large and less hooked than those of other groups such as Phorusrhacidae. Thus, these phalanges are more similar to small hooves than to actual claws, as they present no sharp point (Figure 7.9(F)).

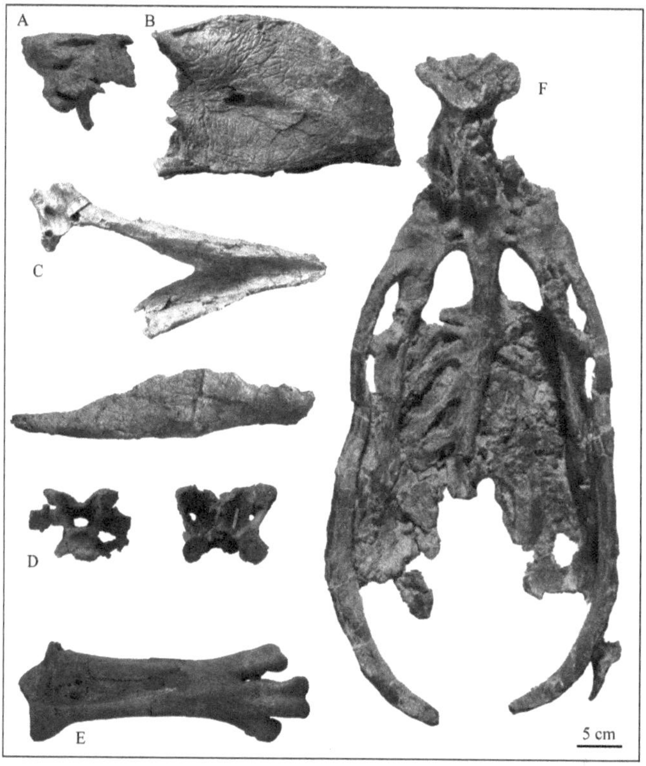

Figure 7.9. *Main skeletal elements of Gastornithidae. (A) Posterior part of skull in lateral view (n° AMNH. 6169), (B) Maxilla in lateral view (n° AMNH. 6169), (C) Mandible in dorsal view (above) and lateral view (below) (n° MHNT. PAL. 1.1.), (D) Vertebrae (left n° MHNT. PAL. 2013.15 in caudal view; right n° MHNT. PAL. 2013.15.6. in dorsal view), (E) Tarsometatarsus in dorsal view (unnumbered, Escuillié collection; cast n° L3092 MNHN), (F) Pelvis in ventral view (n° AMNH. 6169). For a color version of the figure, see www.iste.co.uk/angst/birds.zip*

7.5. Paleophysiology and paleoecology

7.5.1. *Diet*

The assessment of the diet of *Gastornis* started a debate that has been going on for more than 20 years. In 1991, Witmer and Rose carried out the first biomechanical study of the skull and jaw of this bird, concluding that it was a ferocious predator. A year later, in 1992, Andors carried out a similar study on the same fossils and concluded that the diet of this bird was herbivorous. The contradiction between the conclusions of these two studies, carried out nearly at the same time on the same fossil material, besides relying on very similar approaches, shows the limitations of classic comparative methods in paleontology. This problem results from the fact that the skull and mandible of *Gastornis* are anatomically unique and resemble no extant bird (Figure 7.10). Therefore, it is difficult to make a simple anatomical comparison with present-day birds with a known diet.

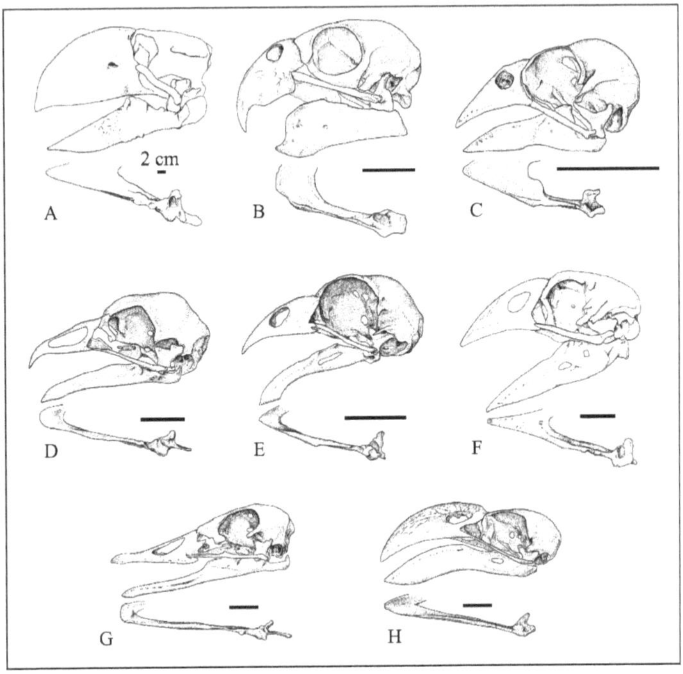

Figure 7.10. *Comparison between the skull and mandible of* Gastornis *and those of extant birds (A) Gastornithidae, (B) Yellow-crowned amazon,* Amazona ochrocephala *(herbivorous, granivorous), (C) Evening grosbeak,* Coccothraustes vespertinus *(herbivorous, granivorous), (D) Southern screamer,* Chauna torquata *(herbivorous, folivorous), (E) Hoatzin,* Opisthocomus hoazin *(herbivorous, folivorous), (F) South Island takahē,* Porphyrio mantelli *(herbivorous, folivorous), (G) Canada goose,* Branta canadensis *(herbivorous, folivorous), (H) Thick-billed raven,* Corvus crassirostris *(carnivorous, scavenger). Re-drawn from Witmer and Rose (1991)*

This debate was settled only in 2014 thanks to the work carried out by Angst *et al.* (2014a) following a multidisciplinary approach that combined isotopic geochemistry and functional morphology. These authors studied the carbon isotopes from the apatite of the bones of *Gastornis* found in French fossil sites in comparison with the carbon isotopes of the fauna associated with this bird at the same sites. They thus were able to show that if this bird was carnivorous, it must have fed on animals whose flesh had a $\delta^{13}C$ equal to −16.3 ± 2.1 ‰. In order to have muscles with such a value of $\delta^{13}C$, these animals must have fed only on C4 plants, the oldest specimens of which are known only from the Eocene–Oligocene boundary, 14 Ma later. On the contrary, if *Gastornis* was a herbivore, it must have fed on plants with a $\delta^{13}C$ ranging between –24.2 ± 2.1 ‰ and –21 ± 2.1 ‰, which corresponds to the range of values known today for C3 plants, which must have formed the life environment of this bird. Furthermore, this range of values is comparable to the range of $\delta^{13}C$ of the plants that were part of the life environment of *Gastornis*, estimated on the basis of the herbivorous mammals known from the same fossil sites. Therefore, this first approach allows us to conclude that the Gastornithidae were herbivores (Figure 7.11).

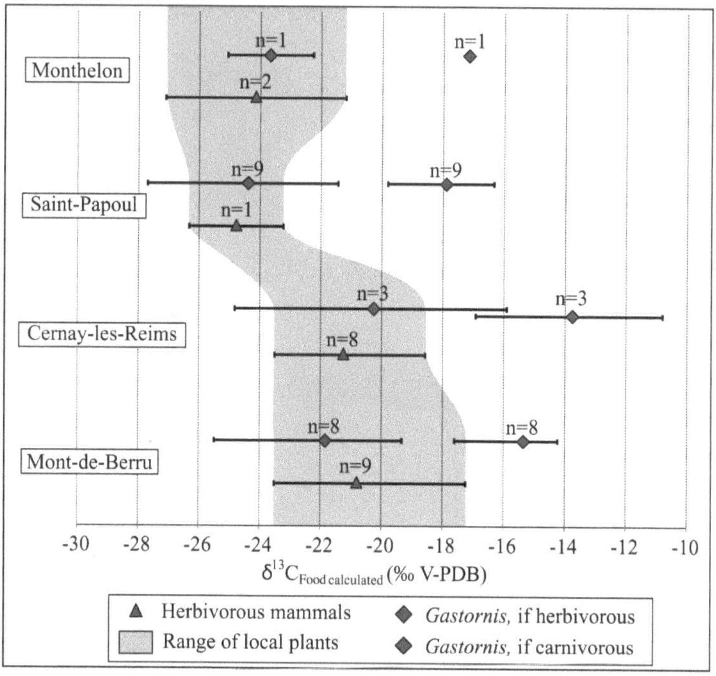

Figure 7.11. *$\delta^{13}C$ of the diet of* Gastornis *and contemporaneous herbivorous mammals from the same fossil sites, estimated from the $\delta^{13}C$ of the apatite of the carbonates for the four sites studied. Range of $\delta^{13}C$ of local plants estimated based on the $\delta^{13}C$ of the fauna of herbivorous mammals from these sites. Horizontal lines: minimum and maximum values, triangle and diamond: mean value, cross-hatched area: boundary between different sites, n: number of specimens. (from Angst* et al.*, 2014a). For a color version of the figure, see www.iste.co.uk/angst/birds.zip*

The second method used to assess the diet of *Gastornis* involves a study of the functional morphology of the mandible of this bird. By studying the musculature of various skulls and mandibles of extant birds with a known diet, it was possible to establish a model (see Introduction). Thus, an herbivorous bird has a highly developed adductor muscle for the mandible and a very reduced pterygoid muscle. The mandibles belonging to the individuals of *Gastornis* kept at the American Museum of National History in New York and at the Toulouse Natural History Museum are especially well preserved and show the traces of the areas where these muscles were attached. Thus, it has been possible to demonstrate that *Gastornis* had a typical musculature of herbivorous birds (Figure 7.2). This second approach, which is completely independent of the first one, confirms that the diet of this bird was herbivorous.

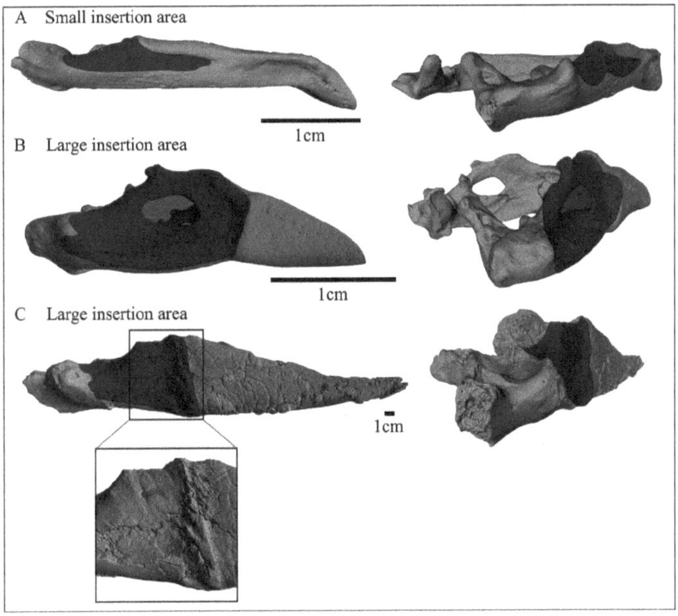

Figure 7.12. *Comparison between the insertion area of the adductor muscle of the mandible (dark area) in extant birds and* Gastornis *in lateral view (left side) and oblique view (right side). (A) Extant carnivorous bird (*Buteo platypterus*), (B) Extant granivorous passerine (*Geospiza magnirostris*), and (C)* Gastornis *mandible and a close-up on the muscle insertion area, allowing an estimation of the size of the adductor muscle of the mandible of this fossil bird (from Angst* et al.*, 2014a)*

Moreover, during field campaigns in southern France, small quartz pebbles that are smooth and opaque, either white or beige, ranging in size from about 0.5 to 1 cm, with a perfectly rounded surface, have been found nearly everywhere in association with the fossil eggshells attributed to Gastornithidae (Figure 7.13). At all the sites, no surrounding rock could have produced these small pebbles by erosion. Therefore, Cailleux (1969a, 1969b) suggested in 1969 that they may be gastroliths, that is, small pebbles swallowed by Gastornithidae to help them digest plants in their gizzard, as is the case for present-day ostriches, whose gastroliths are comparable in size with the small pebbles found together with the fossil eggshells of *Gastornis* (Wings, 2004). This hypothesis is backed up by several lines of evidence. Present-day birds, when choosing their gastroliths, prefer oblong pebbles to round stones, light-colored – for example, yellow, light green or white – and made of silicates (Best, 1995). The small pebbles found at these sites meet all these conditions. If they are actually gastroliths, this would suggest that Gastornithidae had the same digestive physiology as some present-day herbivorous birds, with a gizzard containing gastroliths allowing them to crush more effectively the ingested plants. Nevertheless, it is difficult to test this hypothesis once and for all, and the presence of these small pebbles may be merely due to chance or the presence at these sites of other birds, either contemporaneous with the Gastornithidae or still alive today, as these pebbles are found on the surface together with eggshells.

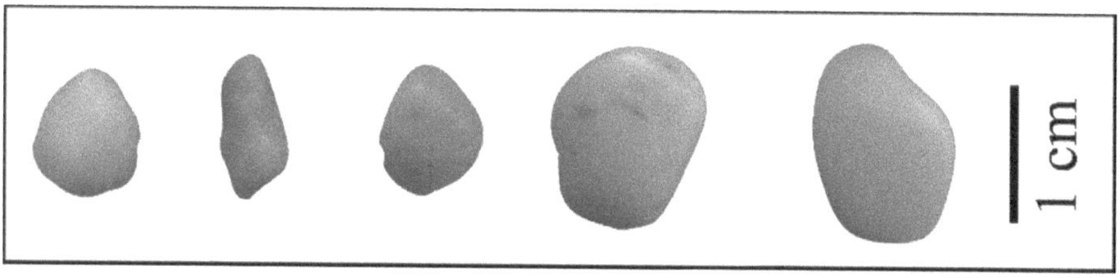

Figure 7.13. *Pebbles that may correspond to gastroliths associated with Gastornithidae*

7.5.2. *Sexual dimorphism*

Before 2015 and the study carried out by Mourer-Chauviré and Bourdon (2015), no sexual dimorphism had ever been suggested for Gastornithidae. The idea of these authors is to account for the wide differences in size observed for the tarsometatarsus and the terminal phalanges of *Gastornis* sp. found in Louvois with a possible form of sexual dimorphism, ranging from

29% to 65% between the largest and the smallest specimens depending on the bones and the measurements considered. This type of sexual dimorphism has been shown so far only concerning the specimens from Louvois, which are provisionally referred to *Gastornis* sp. but probably do not belong to the same species as the other Gastornithidae described so far from Europe (Mourer-Chauviré & Bourdon, 2015). Whether the largest individuals are male or female remains uncertain. On the contrary, no sexual dimorphism has been identified in the other species of European, North American or Asian Gastornithidae.

7.5.3. *Reproduction*

7.5.3.1. *Identification of the eggs*

It has been first of all necessary to identify the eggs of Gastornithidae in order to study how they reproduced. This was possible thanks to the very large number (more than 1300) of eggshell fragments collected in southern France. The vast majority of these eggshells (95%) correspond to the *Ornitholithus arcuatus* type. The eggshells of *O. arcuatus* are the thickest sampled (2 mm, in comparison with 1 mm for the second type of eggshell called *O. biroi*) and occur in the Ypresian in southern France and in the Thanetian in northern Spain (see above). Using Williams' method (1981), which relies on measuring the curvature of the fragments of eggshells (see Introduction), it has been possible to demonstrate that these fragments belonged to eggs that were on average 12 cm wide (range: 11–13 cm; Figure 7.14) and 17.8 cm long (range: 16.3–19.2 cm) (Angst *et al.*, 2014b).

Thanks to the equations published by Hoyt (1979), it was possible to estimate that these eggs had an average volume of 1330.4 cm^3 (range: 1005.9–1654.8 cm^3), which corresponds to a mass of 1.4 kg (range: 1.1–1.8 kg) (Angst *et al.*, 2014b). An egg of this size and mass corresponds to the egg of a present-day ostrich, and, consequently, it was laid by a large bird. In southern France and at this period, the only birds known to be large enough to have laid these eggs are Gastornithidae. By using the equation put forward by Dyke and Kaiser (2010), on the basis of the mass of the egg, it is possible to estimate the length of the femur of the bird that laid it, and the value obtained is very close to the lengths of the known femora of Gastornithidae (Figure 7.15). Thus, it has been possible to refer the eggshells of the *O. arcuatus* type to Gastornithidae (Angst *et al.*, 2014b).

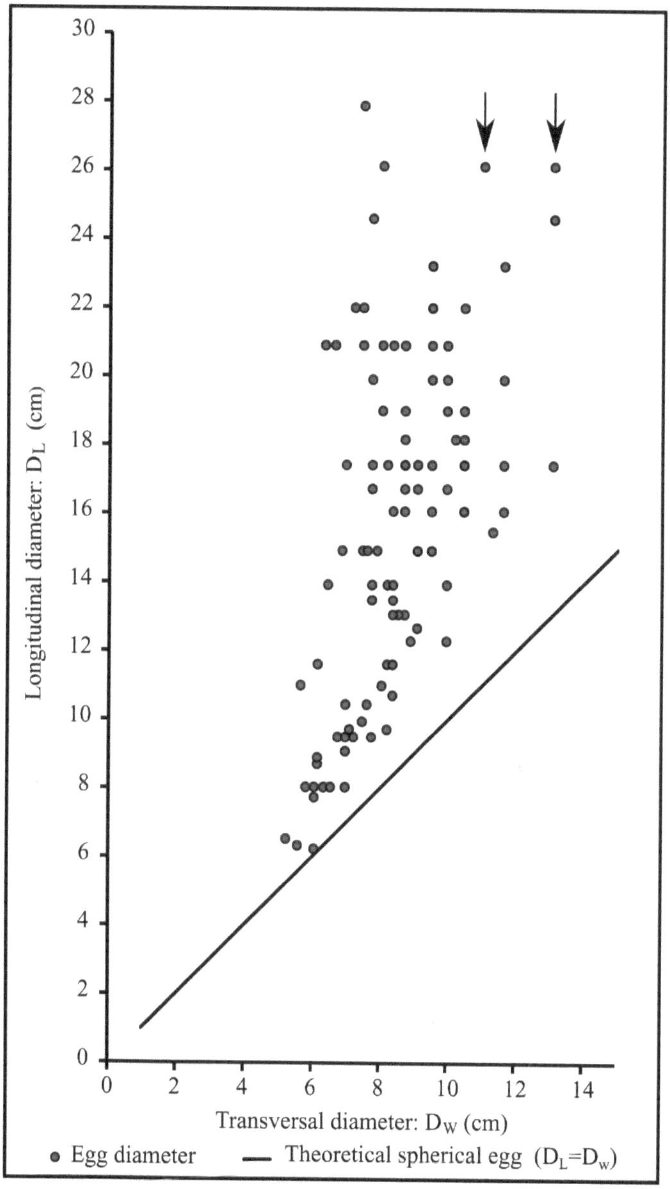

Figure 7.14. *Estimate of the width of the egg obtained using Williams' graph (1981), which represents the transversal diameter (D_W) in relation to the longitudinal diameter (D_L) in centimeters. The two values indicated by the arrows (in the upper right-hand corner) correspond to the eggshells with the largest D_L and D_W, namely the flattest in all spatial directions. The D_L values of these two points will be used as an estimate of the width of the egg*

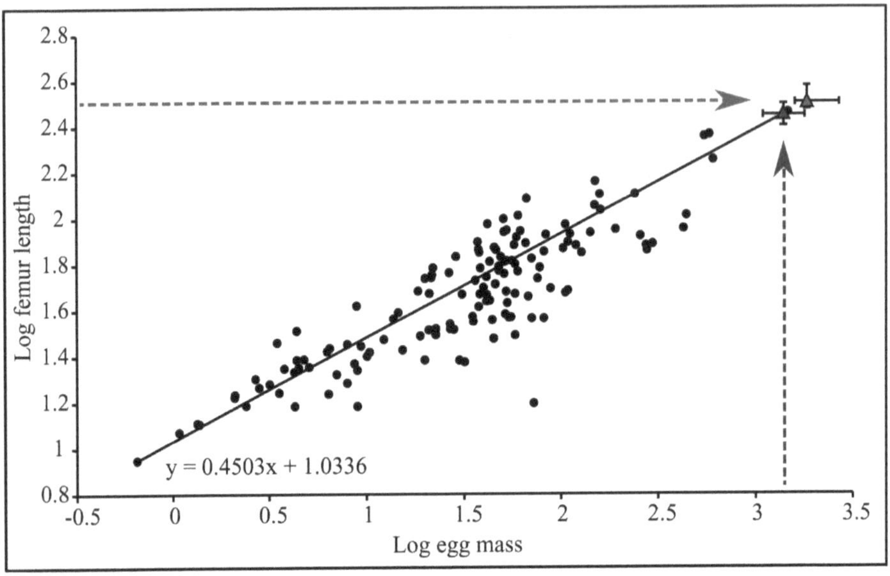

Figure 7.15. *Estimate of the log(femur length) based on the log(egg mass) obtained using the measurements and the regression line published by Dyke and Kaiser (2010) (bottom triangle); compared with the log(egg mass) estimated on the basis of the log(femur length) measured for European* Gastornis *(top triangle). The two values are very similar, suggesting that the fossil eggshells may be attributed to* Gastornis. *For a color version of the figure, see www.iste.co.uk/angst/birds.zip*

By attributing these eggshells to Gastornithidae, it is possible to obtain a substantial amount of information about how these large land birds reproduced.

7.5.3.2. *Nesting*

It should first be noted that no entire egg has been found at the sites where the eggshells have been sampled. Only Dughi and Sirugue (1959), as well as Touraine (1960), mentioned sub-complete eggs discovered in southern France. However, as these eggs have been neither illustrated nor collected, and because no new sub-complete egg has been discovered during the several prospecting campaigns carried out over the last few years, it is difficult to test whether their reports are reliable or not. Therefore, we can assume that nearly all the eggs of Gastornithidae laid in southern France were destroyed either before or during fossilization. In southern France, a large number of complete dinosaur eggs have been discovered in the Cretacean strata since the mid-19th Century (Pouech, 1859). The main hypothesis put forward to account for the excellent preservation of such a large number of dinosaur eggs is that these eggs were laid in holes and then partly buried so that entire eggs were likelier to become fossilized (Erben *et al.*, 1979; Kérourio, 1981; Horner, 2000; Paik *et al.*, 2004). By comparison with these particularly well-preserved eggs, the opposite hypothesis may be put forward: the eggs of Gastornithidae must have been laid on the ground (as is the case for present-day ostriches) without being buried in any way, which did not allow a good preservation, which would in turn explain why only fragments of eggshells can be found.

7.5.3.3. *Egg-laying strategy*

Another observation about these eggshells may provide more information about the egg-laying strategy of these birds. The external surface of the eggshells presents a variable ornamentation, which may correspond to a wear gradient of the shell. This hypothesis is supported by the study of some shell fragments that show several types of progressively eroded ornamentation (Figure 7.16(G)). The abrasion of these shells may have two causes: a taphonomic cause or an ecological cause. In case of wear associated with the fossilization process, because of the action of microorganisms or of the abrasion linked to the circulation of fluids, the shell surface would have been modified both internally and externally. However, there is no noticeable wear on the inner surface of the shells. Therefore, it may be assumed that this different ornamentation has been caused by an abrasive process when the eggs were still whole, while they were being laid and before hatching, so that the external surface was affected. Similar observations can be made about the surface of the eggs of present-day ostriches, which present several ornamentations depending on their position on the egg. These variations are due to the friction between eggs in the nest. By comparison, it can be assumed that gastornithids laid several eggs in the same nest. This may be the result of two types of egg-laying strategies: either each female laid several eggs at once in the same nest, as is the case for present-day ostriches, or the same nest was shared by several females, each of which laid only one egg, as is the case for present-day birds such as Acorn woodpeckers (*Melanerpes formicivorous*) (Danchin *et al.*, 2005, p. 410), so that they can watch over the nest in turn and allow the other adults to feed.

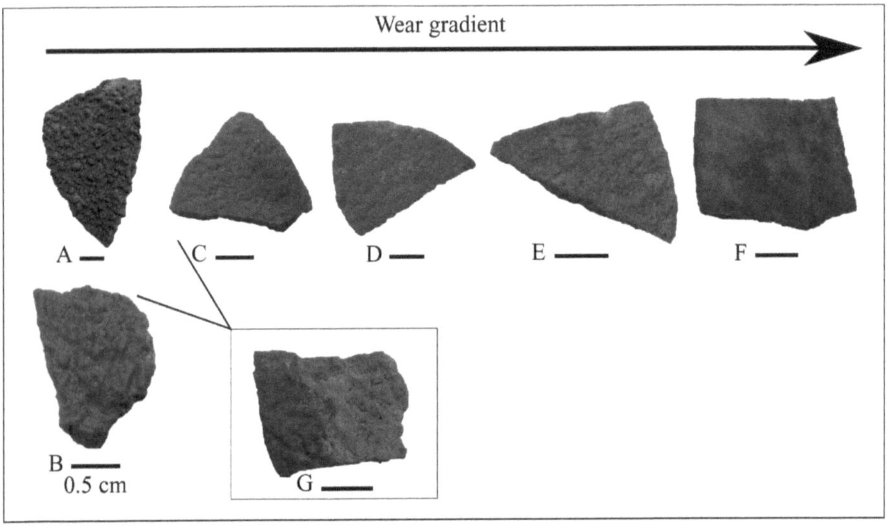

Figure 7.16. *Variation of the external ornamentation on the fossil eggshells of* O. arcuatus *in relation to wear. (A) Mamillated ornamentation, (B) Ornamentation with marked vermiculation, (C) Ornamentation with moderate vermiculation, (D) Ornamentation with fine and spaced-out vermiculation, (E) Semi-smooth ornamentation, (F) Smooth ornamentation, (G) Shells showing the transition between ornamentations B and C. For a color version of the figure, see www.iste.co.uk/angst/birds.zip*

7.5.3.4. *Autonomy of the hatchlings*

These eggshells also provide information about the females that laid the eggs. It is possible to estimate the body mass of the female on the basis of the mass of the eggs. Thanks to the egg fragments of Gastornithidae, it was possible to obtain the volume of an entire egg by using the curvature of the eggshells (see Introduction, Angst *et al.*, 2014b) and, on the basis of this estimate, it was possible to calculate the mass of the eggs (see above). To determine the body mass of the females that laid these eggs, there are several equations related to the precociousness of the hatchlings (see Introduction, Dyke & Kaiser, 2010; Angst *et al.*, 2014b). Either Gastornithidae were birds whose hatchlings were independent from birth (precocial) or these hatchlings needed to be taken care of by their parents after birth (altricial). In the former case, the female *Gastornis* in southern France would have had a body mass ranging from 135 to 156 kg, whereas in the latter case, their mass would have ranged from 67 to 80 kg. No sexual dimorphism has been demonstrated so far in Gastornithidae, apart from the species found at Louvois, which is probably not the one that laid the eggs found in southern France. Furthermore, the body mass estimates obtained with other methods and on a wide range of bone remains show that the body mass of French Gastornithidae (apart from the species found at Louvois) ranged from 108.1 to 180.2 kg (Angst *et al.*, 2014b). These values are very likely to include both male and female individuals, as it seems difficult to explain why fossilization would favor a sex rather than the other. Consequently, this excludes the possibility that females had a body mass that ranged between 67 and 80 kg, which is too different from the masses estimated for the rest of the population. Therefore, this provides another indirect piece of information about the ecology of Gastornithidae, whose hatchlings became autonomous right after birth (precocial), as is the case for present-day ostriches (Dyke & Kaiser, 2010).

7.5.3.5. *Egg-laying sites*

During several prospecting campaigns in southern France, even though no entire eggs were found, a large number of eggshells referred to *Gastornis* were collected. The significant accumulation of egg fragments may indicate the presence of preferential egg-laying sites. It is undoubtedly difficult to get a precise grasp of the chronological succession of the events at different sites, as the stratigraphy is not accurate enough. Nevertheless, the significant accumulation of eggshells may result from not only several eggs laid simultaneously by different females but also a series of eggs laid one after the other over time and over several years. The only certain thing is that such a concentration of shells cannot derive from a single nest. Therefore, this observation indicates that *Gastornis* in southern France laid their eggs at particular sites, probably because the environmental conditions in those places were more favorable. If we admit that these accumulations of eggshells result from not only the presence of several nests built successively over the years but also several nests that belonged to different females, this would suggest that these large birds lived in groups, at least during reproduction, as is the case for present-day ostriches (Bertram, 1980). However, this hypothesis is hard to test without a finer stratigraphy, which would allow us to see whether the shells derive from eggs laid simultaneously or not.

7.5.3.6. *Egg-laying period*

The last piece of information that can be obtained from these fossil eggshells belonging to *Gastornis* is the period during which these birds laid their eggs. This information was obtained thanks to geochemical analyses of the carbon and oxygen isotopes of these shells. The $\delta^{13}C$ and $\delta^{18}O$ provide information about rainfall and average air temperature during the formation of the eggshell, respectively (see Introduction). The isotopic measurements allow us to estimate the annual rainfall at about 500 ± 300 mm per year and an average yearly air temperature that ranged between 8 and 22°C. Thus, these values show that the environment was hot and dry, as recorded by the eggshells during their formation and confirmed by the measurements of the porosity of the eggshells obtained by Donaire and Lopez-Martinez (2009). In relation to the egg-laying strategy of Gastornithidae, these estimates may reflect:

– Either the annual climate, which was hot and dry year-round if we consider that these large birds laid eggs all year long, as is the case for some present-day tropical birds (e.g. *Columbina passerina*, Campbell & Lack, 1985)

– Or the climate only during the egg-laying period, which would then differ from the annual climate if these birds laid eggs only at a specific period. For example, present-day ostriches lay eggs only at the beginning of the humid season (Deeming, 1999).

The reconstruction of the climate during the Paleocene-Eocene in northern Spain, put forward by Schmitz and Pujalte (2007), shows periods with heavy rainfall that cannot be detected in the fossil record. Two hypotheses may account for this difference. The hydric regime may have increased markedly during the PETM, being thus recorded by Schmitz and

Pujalte's study, which focuses on this period, but not by the eggshells, which are older or more recent than the PETM. Therefore, the climate in southern France before and after the PETM would have been more arid than during this period, as well as more arid than in northern Europe, where the environments known and described show heavier rainfall (e.g. 2091 mm per year at the Geiseltal, Teodoridis *et al.*, 2012). The second possible hypothesis is that Gastornithidae did not lay eggs all year round but only in a drier period, that is, just before the humid season, as is the case for present-day ostriches. In this case, the fossil eggshells would have recorded only the climate of this part of the year. If the latter hypothesis is correct, we may wonder how many dry periods there were in a year during which Gastornithidae could lay eggs. If we compare the rainfall and the temperature values obtained thanks to the eggshells with the present climate, considering that they correspond to the values of the dry period, the only compatible climate is the Mediterranean climate, such as that of the Balearic Islands, which has only one dry period in summer. By comparison, it may be suggested that Gastornithidae only laid eggs at the end of the summer, just before the beginning of the autumnal rainfall, and that the climate in southern France was Mediterranean in type.

7.5.4. *Locomotion*

Gastornithidae had a relatively slow and graviportal type of locomotion, as shown by the method of Angst *et al.* (2016) and the approach adopted by Storer (1960). The tarsometatarsus is relatively short and its maximum length/minimum width ratio ranges from 5 to 8 (Figure 7.17), which is clearly less than the ratio of 12 that marks the boundary between graviportal and cursorial (Angst *et al.*, 2016). This type of locomotion had already been suggested on the basis of indirect elements, such as the footprints attributed to *Gastornis*. These footprints do not show any significant relation to the footprints of herbivorous mammals that could indicate a potential predator/prey link between Gastornithidae and small contemporaneous mammals whose footprints are preserved at the same sites, so that there is no evidence that they rapidly chased their prey (Mustoe *et al.*, 2012). Furthermore, all Gastornithidae had the same type of locomotion regardless of their geographical and stratigraphic origin or their body mass (North American specimens are larger and heavier than the European species).

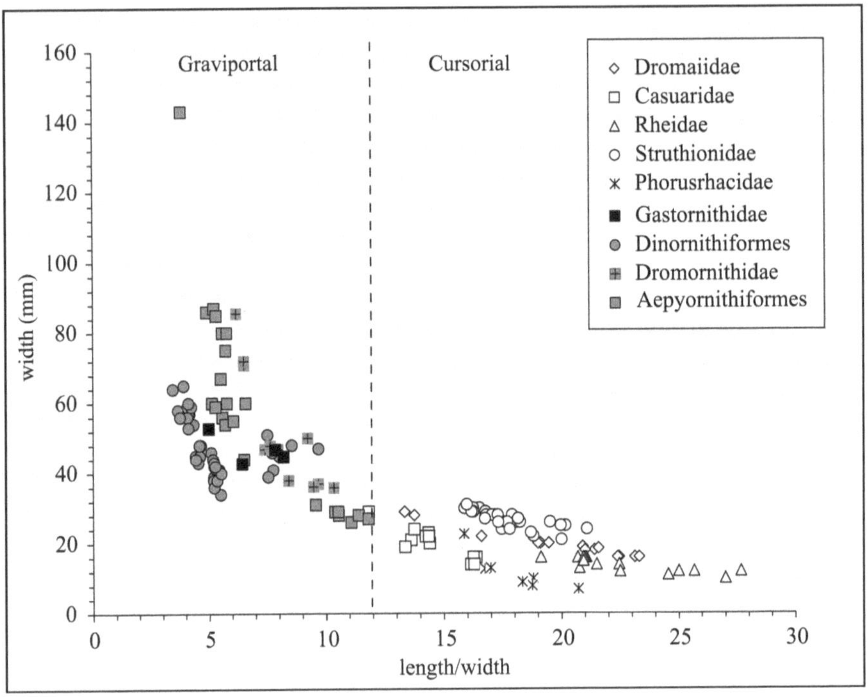

Figure 7.17. *Estimate of the locomotion type of Gastornithidae based on the method proposed by Angst* et al. *(2016)*

7.5.5. *Body mass*

7.5.5.1. *Body mass of European Gastornithidae*

Several estimates of the body mass of Gastornithidae have been suggested since the 19th Century, starting with the value of 500 kg put forward by Hébert (1855b) for *Gastornis parisiensis* by comparing *Gastornis* with an albatross (Table 7.1). At the same time, Prévost (1855a) put forward a body mass of nearly 200 kg by extrapolation based on the bones of a swan. The only recent estimates are those put forward by Angst *et al.* (2014b), who have obtained average values that range from 108 to 180 kg by using the Campbell and Marcus equations (1992). On the basis of the relation between the size of the eggs and the mass of the female that laid them (Dyke & Kaiser, 2010), it was possible to determine the body mass of female European *Gastornis* (Angst *et al.*, 2014b). It ranges from 135 to 156 kg in case these birds were precocial (i.e. hatchlings that become rapidly autonomous after birth) or from 66 to 79 kg if their ecology was altricial (hatchlings that need parental care after birth). The lowest values (66–79 kg) are very low for a bird as large as Gastornithidae, and they could be explained only by the presence of significant sexual dimorphism, whereby females were much smaller and lighter than males and their size was halfway between that of present-day ostriches and emus. If we considered the second hypothesis to be correct, this would mean that the only fossils hitherto referred to Gastornithidae (apart from the specimens from Louvois) belong to male individuals. Even though the fossil record of this group is relatively limited, it seems difficult to explain such a selective preservation between male and female *Gastornis*. Consequently, it seems more correct to think that these birds were not characterized by a pronounced sexual dimorphism (excluding the *Gastornis* from Louvois, for which no body mass estimate is available), thus concluding that their ecology was precocial, as in present-day ratites. After integrating additional bone measurements in Campbell and Marcus's models, the body masses obtained for the European forms are slightly more scattered than those put forward by Angst *et al.* (2014b), ranging from 108 to 193 kg instead of varying from 108 to 180 kg.

The body mass of the second European species of *Gastornis*, *G. geiselensis*, is 180 kg. Nevertheless, the very limited amount of material does not allow us to assess any potential intraspecific variation, making it difficult to compare it with the other species of *Gastornis*. Finally, as no femur or tibiotarsus is known for the last two European species of Gastornithidae, *G. russelli* and *G. sarasini*, no estimate of their body mass has been put forward so far. The same problem arises with the potential new species of *Gastornis* from Louvois.

Method	Campbell & Marcus (1992) (femur) (kg)	Campbell & Marcus (1992) (tibiotarsus) (kg)	Campbell & Tonni (1983) (tibiotarsus) (kg)	Hurlburt (1999) (GDI model) (kg)	Dyke & Kaiser (2010) (egg) (kg)	Other or unspecified (kg)
Gastornis parisiensis	156 (122–180) [2-6]	133 (108–62) [2–6] 149 (108–193) [7]			135–156 [2] 66–79 [2]	500 [3] 200 [5]
Gastornis geiselensis	180 [6]					
Gastornis giganteus	219–259 [4] 210 (183–229) [6]	180 (160–190) [6]	175 [1]	199 [4]		

Table 7.1. *Compilation of the body masses estimated for Gastornithidae in the literature. Average value (minimum value – maximum value); [1]: Andors (1992), [2]: Angst* et al. *(2014b), [3]: Hébert (1855b), [4]: Murray & Vickers-Rich (2004), [5]: Prévost (1855a), [6]: Angst (2014), [7]: pers. obs.*

7.5.5.2. *Body mass of North American Gastornithidae*

Andors (1992) was the first to put forward a body mass of 175 kg for North American Gastornithidae, employing the method used by Campbell and Tonni (1983) that relies on the minimum circumference of the tibiotarsus (Table 7.1). Later, Murray and Vickers-Rich (2004) used several methods to re-estimate this mass. Thanks to the graphic double integration (GDI) approach, they obtained a value of 199 kg for *G. giganteus*, without providing any information about the parameters they used to build their restoration of the body. They also used the method proposed by Campbell and Tonni (1983) as well as Andors (1992), obtaining a mass that ranges from 215 to 259 kg (Table 7.1). Nevertheless, the masses estimated by Murray and Vickers-Rich (2004) cannot be used, as the authors made a conversion mistake when transforming mathematically the Campbell and Marcus equations (1992). Therefore, all their estimates based on these equations are wrong, as Nguyen *et al.* (2010) have already demonstrated. Furthermore, they mixed several circumference measurements in some cases, without always applying the right measurement to the right equation. The most recent estimate of the body mass of *Gastornis giganteus* is the one made by Angst (2014) and calculated thanks to the Campbell and Marcus equations (1992). The minimum circumferences of the femur yield a mass that ranges from 183 to 229 kg (an average of 210 kg) and, on the basis of the minimum circumferences of the tibiotarsus, the mass ranges from 160 to 190 kg (an average of 180 kg). Therefore, the mass of *G. giganteus* must have ranged from 160 to 229 kg, which is significantly more than what has been estimated for the European species. This difference may be directly linked to the difference in size between these birds on both sides of the Atlantic. *Gastornis giganteus* is about 25% larger than *Gastornis parisiensis*. This difference can be seen in particular in their mandible (Angst & Buffetaut, 2013). *G. giganteus* was also about 25% heavier than *G. parisiensis*.

7.6. Extinction

The first wave of extinctions of Gastornithidae took place in North America at the end of the Ypresian (MP10), that is, more than 5 Ma before the definitive disappearance of this group in Europe. Nevertheless, the causes behind their extinction remain unknown.

The last body fossils of Gastornithidae are from the Middle Eocene (MP13) in Europe, at the Geiseltal site in Germany. The causes behind the extinction of this group are not clear yet, but there are several hypotheses.

7.6.1. *Climate and environmental changes*

The first factor that may have caused the disappearance of Gastornithidae is a significant climate change, which could have affected communities as well as the sea level, which could have rapidly fragmented the distribution area of Gastornithidae. However, the paleoclimatic reconstructions for the Eocene in Europe do not reveal any significant and brutal change in the temperature of deep waters or in the mean sea level (Mosbrugger *et al.*, 2005; Sluijs *et al.*, 2008; Zachos *et al.*, 2008; Miller, 2009; Utescher *et al.*, 2009). Instead, the climate changed progressively from the PETM onward, with gradually decreasing temperatures and sea levels (Figure 7.18). These climate changes led to the evolution of the paleoenvironments in Eastern Europe, with tropical and evergreen forests at the beginning of the Lutetian in Germany (the Geiseltal) replaced by tropical-to-subtropical evergreen forests associated with deciduous trees that lost their leaves seasonally at the end of the Lutetian in Germany (Scheiplitz) (Utescher *et al.*, 2009) (Figure 7.18). At the same time, the sea level sunk gradually, without probably affecting the area of distribution of Gastornithidae. On the contrary, this probably facilitated faunal exchanges with the rest of Europe. The climate factors and the evolution of the flora and the paleoenvironments, therefore, cannot on their own account for the disappearance of Gastornithidae in the middle Lutetian.

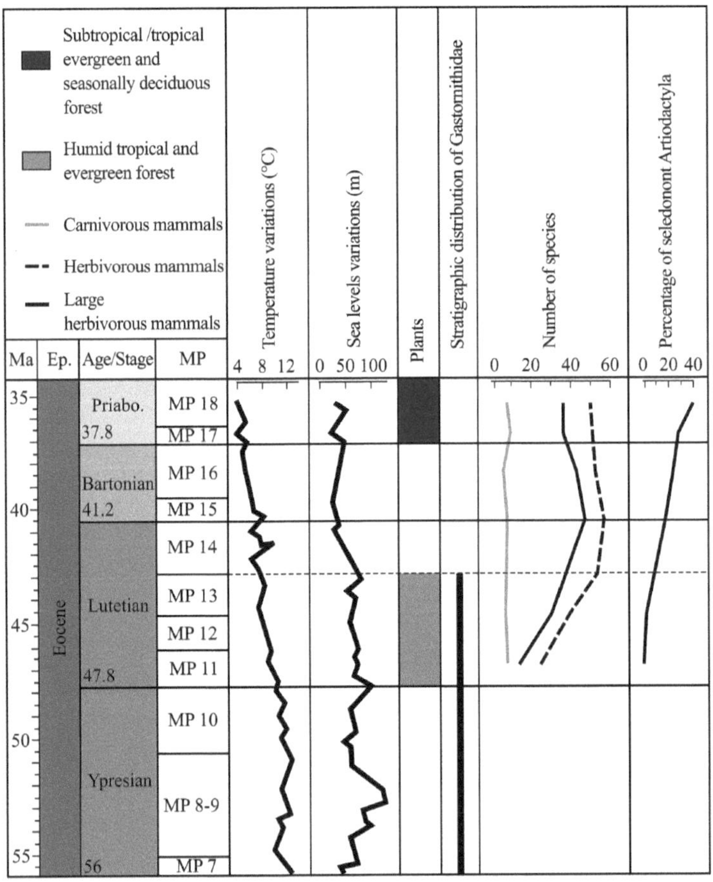

Figure 7.18. *Variations of the environmental conditions and in the number of mammal species, and the stratigraphic distribution of Gastornithidae in Europe during the Eocene. Temperature variations (from Zachos* et al.*, 2008), sea level variations (from Miller, 2009), changes in the vegetation (from Utescher* et al.*, 2009), number of species (from Legendre* et al.*, 1991, and Sudre & Legendre, 1992) and the percentage of selenodont Artiodactyla (from Sudre & Legendre, 1992)*

7.6.2. *Changes in the mammal fauna*

The second possible cause for the extinction of Gastornithidae could be the evolution of the land mammal fauna. It might be assumed that large carnivorous mammals appeared at the end of the Lutetian and preyed on *Gastornis*. Another hypothesis involves the evolution of herbivorous mammals that may have started competing with *Gastornis* for access to food.

The first possibility may involve either the significant diversification of carnivorous mammals of all sizes or the appearance of at least a large carnivorous species. No diversification of carnivorous mammals is known to have taken place during the Middle Lutetian in Europe, as the number of species remained relatively steady until the end of the Middle Eocene (Figure 7.18; Legendre *et al.*, 1991). There are only two groups among these carnivores whose sizes are smaller than that of a small feline, namely Carnivora and Creodonta (Solé *et al.*, 2015). The second hypothesis involves the diversification of herbivorous mammals, which may have been more specialized than Gastornithidae and replaced them in specific ecological niches. They may have also been larger and replaced *Gastornis* in the ecological niche occupied by the large generalist herbivores. The studies carried out by Legendre *et al.* (1991), as well as those conducted by Sudre and Legendre (1992), partially answer this question. These authors noted a significant diversification among herbivorous mammals from the beginning of the Lutetian, which became more pronounced in MP 12-13, that is, around the same time as the extinction of *Gastornis* (Figure 7.18). Simultaneously, Sudre and Legendre (1992) highlighted the appearance of artiodactyl herbivorous mammals with selenodont teeth with the crescent-shaped dental cusps that characterize present-day ruminants (Figure 7.18). These two observations demonstrate that the Lutetian corresponded to a period of specialization for European herbivorous mammals, which might have also colonized specific ecological niches in which *Gastornis* might have occasionally fed. Furthermore, Sudre and Legendre (1992) reported a significant and simultaneous increase in the number of species of large herbivorous mammals, which may have competed with Gastornithidae (Figure 7.18). Therefore, the diversification of herbivorous mammals, which started at the beginning of the Lutetian, could have progressively led to the extinction of these large birds, as their ecological niches were gradually occupied by more specialized mammals, which were consequently more adapted to each of these specific niches.

Nevertheless, these various hypotheses about the extinction of Gastornithidae must be compared with another alternative concerning the stratigraphic record. These hypotheses become valid only if we regard the stratigraphic record of these birds as relatively complete and reliable. If this record is not complete and these birds became extinct at a later stage, the spatiotemporal framework in which the extinction of this family took place would change.

8

Gargantuavis

8.1. Historical background

The first fossil indicating the existence of a giant bird during the Late Cretaceous in southern France was a fragment of synsacrum discovered at the Bastide Neuve site in Fox-Amphoux (Var, Provence; Buffetaut *et al.*, 1995). The discovery of more complete elements in other sites in southern France (a pelvis at Campagne-sur-Aude, in the Aude department, chosen as the holotype, and an incomplete femur at Villespassans in the Hérault department) then made it possible to erect a new taxon, *Gargantuavis philoinos* Buffetaut & Le Loeuff, 1998. Since then, other skeletal remains have provided a slightly better understanding of this bird (Buffetaut & Angst, 2016a): a cervical vertebra (Buffetaut & Angst, 2013), pelvic elements (Buffetaut & Angst, 2016b) and a femur from Cruzy (Hérault department), two incomplete pelves from Fox-Amphoux (Buffetaut *et al.*, 2015), and a synsacrum from Laño (in Burgos province, Spain; Angst *et al.*, 2017). However, *Gargantuavis philoinos* remains poorly known in many respects; in particular, no remains of its skull and jaws have yet been discovered (Buffetaut & Angst, 2016a).

8.2. Taxonomy, systematics and classification

8.2.1. *Taxonomy*

On the basis of the principle of parsimony, all the remains of giant birds that have been discovered so far in the Upper Cretaceous of France and Spain are referred to a single taxon, *Gargantuavis philoinos*, as none of their features contradict this interpretation. There are noticeable differences in size which, however, do not seem to be beyond the limits of individual or sexual variation.

8.2.2. *Systematics*

Class Aves Linnaeus, 1758

Genus *Gargantuavis* Buffetaut & Le Lœuff, 1998

– *Gargantuavis philoinos* Buffetaut & Le Lœuff, 1998

Stratigraphic occurrence: Upper Cretaceous of southern France and northern Spain.

8.2.3. *Classification*

Even though the avian nature of *Gargantuavis philoinos* has been doubted, and it has been suggested that it should be regarded as a pterosaur (Mayr, 2009), the various bones discovered differ quite significantly from the corresponding elements of pterosaurs, besides showing features that indicate clearly that they belong to birds (Buffetaut & Le Lœuff, 2010). The exact position of *Gargantuavis philoinos* within the group, however, remains uncertain due to the lack of information about many aspects of its anatomy. This taxon exhibits a combination of features which is difficult to interpret systematically. The pelvis presents primitive features and is characterized, in particular, by a synsacrum that includes a reduced number of vertebrae (about ten). The acetabulum is in a very anterior position in relation to the vertebrae that form the synsacrum, undoubtedly because the synsacrum includes only a few thoracic vertebrae. Moreover, the ilia are not fused dorsally above the synsacrum. On the contrary, the only known cervical vertebra has a derived aspect and is markedly heterocoelous, similar to those of modern birds. The well-preserved femur recently discovered at Cruzy looks quite modern, especially because of the lack of a posterior trochanter, which can be found in more basal birds, including Enantiornithes and *Archaeopteryx*, and in Dromaeosauridae. The features of *Gargantuavis philoinos* suggest that this bird was an ornithuromorph more advanced than Enantiornithes, which should possibly be placed in a fairly basal position among Ornithurae, or close to this group. It has been suggested that it may be related to *Patagopteryx*, a land bird from the Late Cretaceous of Argentina (Buffetaut & Le Lœuff, 1998). Nevertheless, *Patagopteryx* seems to be less advanced in some respects, especially in the morphology of the distal part of the femur.

8.3. Stratigraphic and geographical distribution

Gargantuavis philoinos is currently known from five localities, four of which are in southern France (Fox-Amphoux, Campagne-sur-Aude, Villespassans and Cruzy), whereas the other one is in north-western Spain (Laño) (Figures 8.1 and 8.2).

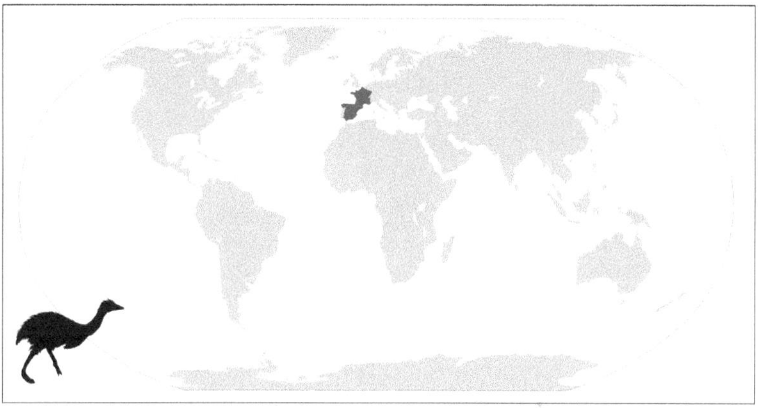

Figure 8.1. *Geographical distribution of* Gargantuavis. *For a color version of the figure, see www.iste.co.uk/angst/birds.zip*

In all instances, its remains come from fluvial deposits. The associated fauna, especially dinosaurs, is typical of the Late Campanian – Early Maastrichtian interval. Magnetostratigraphic studies have provided more precise estimates of the age of two sites that have yielded remains of *Gargantuavis*: terminal Campanian for Laño (Corral *et al.*, 2016) and basal Maastrichtian for Campagne-sur-Aude (Fondevilla *et al.*, 2016). The other sites that have yielded fossils of *Gargantuavis* are likely to date back to the same period, around the Campanian–Maastrichtian boundary at –72 Ma.

From a paleogeographical perspective, the fossil sites containing *Gargantuavis* are part of the Late Cretaceous Ibero-Armorican Island of the European Archipelago, which included a large part of the Iberian Peninsula and present-day France (Figure 8.2).

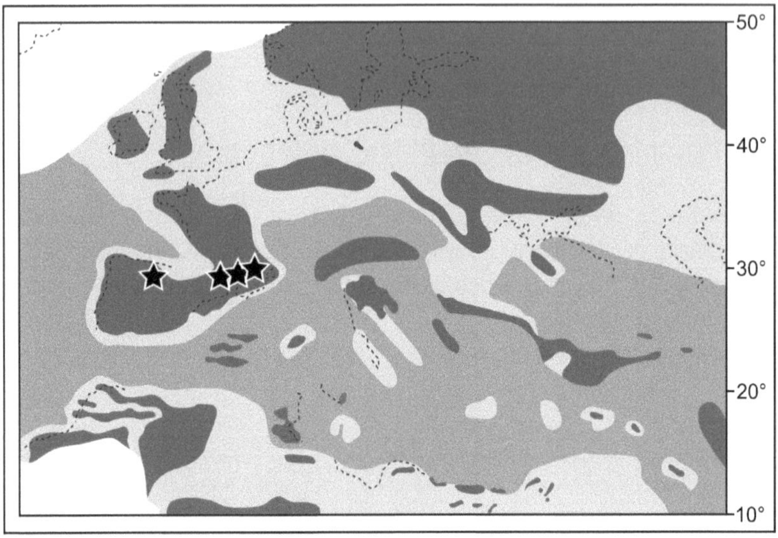

Figure 8.2. *Paleogeographical map of Europe during the Late Cretaceous, showing the distribution of the sites that have yielded* Gargantuavis *(from Buffetaut & Angst, 2016a). For a color version of the figure, see www.iste.co.uk/angst/birds.zip*

8.4. Anatomy

Figure 8.3. *Known skeletal elements and hypothetical restoration of* Gargantuavis *(modeled on* Patagopteryx*)*

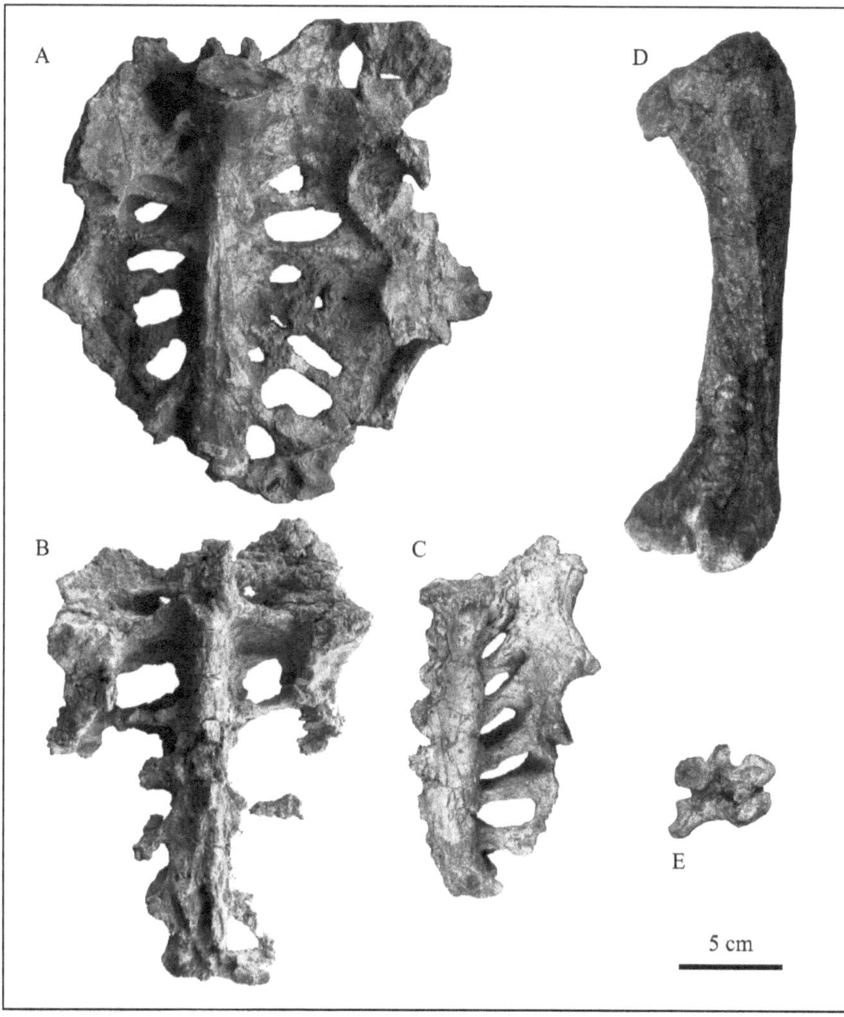

Figure 8.4. *Skeleton elements known to belong to* Gargantuavis philoinos. *(A) Pelvis n° MDE C3-525, (B) Pelvis n° BN 758, (C) Pelvis n° BN 763, (D) Femur n° MN 1335 and (E) Cervical vertebra n° MC-MN476 (photos by Eric Buffetaut). For a color version of the figure, see www.iste.co.uk/angst/birds.zip*

The osteology of *Gargantuavis philoinos* remains incompletely known (Figure 8.3). The only known cervical vertebra is typically avian due to its pronounced heterocoelous condition (Figure 8.4(E)). Nevertheless, it is difficult to determine whether the neck was long or short, based only on this element. The pelvis is remarkable because of the markedly anterior position of the acetabulum and also because the ilia, which form a well-developed anterior blade, are not fused dorsally (Figure 8.4(A–C)). The pelvis seems wide and relatively short. The acetabulum is large, and there is a well-developed antitrochanter in a dorsocaudal position. The femur is robust and relatively short, with a well-individualized articular head separated from the body of the bone by a pronounced neck (Figure 8.4(D)). There is no posterior trochanter, but the trochanteric crest is well developed. The diaphysis is curved. The distal condyles are well individualized with marked reliefs, and the medial condyle is more distally developed than the lateral condyle, which seems uncommon.

A significant feature is the advanced pneumatization of the axial skeleton. The cervical vertebra shows distinct pneumatic foramina. The whole pelvis seems to have been pneumatized, with a vast pneumatic canal running along the synsacrum ventrally to the neural canal, and a significant pneumatization of the ridge formed by the fused neurapophyses of the synsacral vertebrae. The ilium is also hollow, with thin bony walls. On the contrary, no large pneumatic foramina can be identified on the femur.

8.5. Paleophysiology and paleoecology

8.5.1. *Diet*

The absence of data about significant elements of the anatomy of *Gargantuavis philoinos* (especially the skull) limits the possibility of reconstructing the biology of this bird, especially its diet.

8.5.2. *Body mass*

The first body mass estimate made, based on the femur found at Villespassans, provided a value of approximately 140 kg (by comparison with present-day birds), a weight reached by male ostriches (Buffetaut & Le Lœuff, 1998). The femur discovered at Cruzy in 2016, which is much better preserved, is smaller and seems to have belonged to an animal the size of an emu, which might have weighed approximately 57 kilos (an estimate obtained by using the Campbell & Marcus method, 1992). In either case, the estimated masses are much greater than those reached by present-day flying birds. Besides, the wide pelvis and the short and robust femur resemble those of flightless birds, such as some moa.

8.5.3. *Locomotion*

Everything suggests that *Gargantuavis philoinos* was unable to fly. The type of locomotion of *Gargantuavis philoinos* may be deduced to a certain extent from the morphology of its pelvis, which is remarkably wide in relation to its length (Figure 8.4). It differs substantially from the very narrow and elongate pelvis of present-day cursorial ratites such as ostriches, rheas and emus, resembling more the much broader pelvis of some moa, especially graviportal forms such as *Pachyornis* (see Worthy & Holdaway, 2002). Therefore, *Gargantuavis* probably had a graviportal, rather than cursorial, type of locomotion. Nevertheless, since the tarsometatarsus has not been found, it is impossible to apply the methods used for assessing locomotion (Storer, 1960 or Angst *et al*., 2015).

8.5.4. *Bone growth and insularity*

A paleohistological study carried out on the femur from Villespassans (Chinsamy *et al*., 2014) provided information about the type of growth of *Gargantuavis philoinos*. The bone microstructure is avian. Nevertheless, unlike most extant birds (including ratites) whose growth is rapid and uninterrupted until adulthood, it shows that the growth was rapid at first but it was followed by a long period of slow and cyclic growth. This is reminiscent of the extant kiwis (*Apteryx*) and some moa such as *Euryapteryx*, *Anomalopteryx* and *Megalapteryx* (Turvey *et al*., 2005).

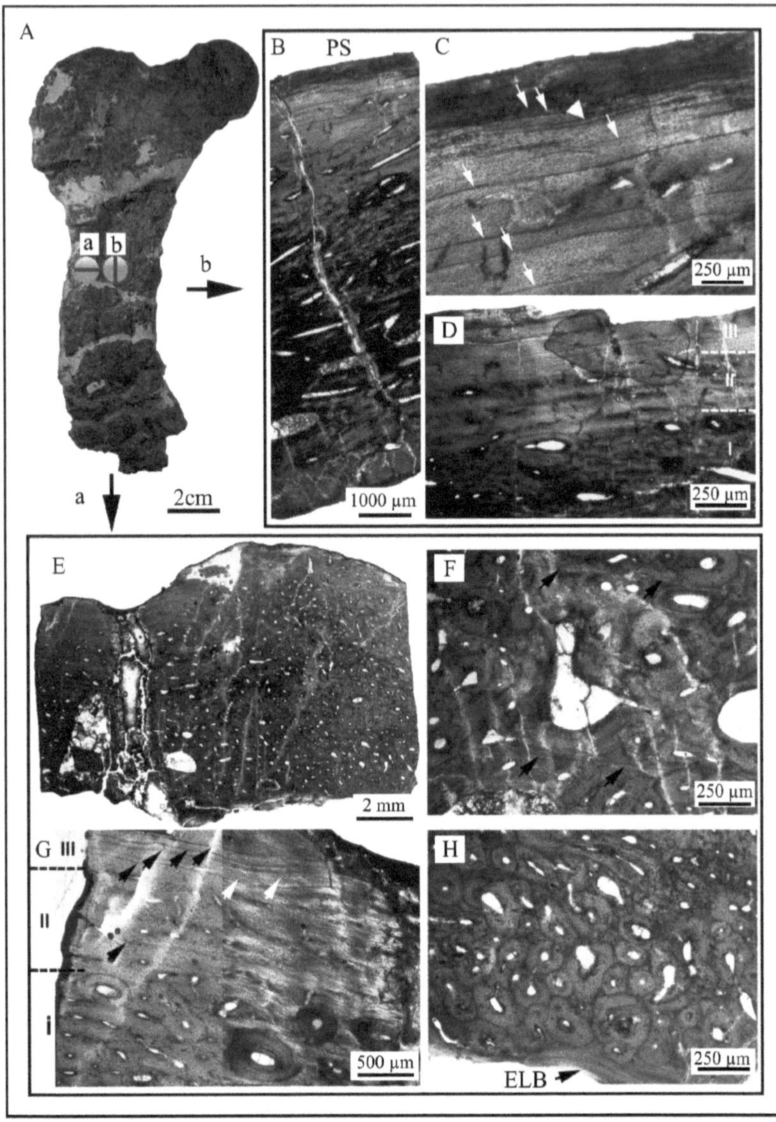

Figure 8.5. *Histology of the femur of* Gargantuavis *(MDE A06). (A) Cranial view of the femur and position of the sites drilled for the histological study, indicated by the gray circles. (B) Part of the longitudinal section of drilling site b. (C) A detail of the outer part of the cortex. (D) A detail of drilling site b showing the change in bone texture. (E) Transversal section of the drilling site. (F, G, H) Details of section a. PS: periosteal surface, ELB: endosteal lamellar bone, White arrows: lines of arrested growth, Thick, white triangle: double line of arrested growth. From Chinsamy* et al. *(2014). For a color version of the figure, see www.iste.co.uk/angst/birds.zip*

The reasons behind this peculiar growth pattern may be found in the insular environment in which *Gargantuavis* lived. This is also the case for the aforementioned insular birds, and also for the titanosaurian dinosaur *Ampelosaurus* (Klein *et al.*, 2012), known from the Upper Cretaceous of France, especially at Campagne-sur-Aude, which is a site that has also yielded a pelvis belonging to *Gargantuavis*. A similar type of growth has also been observed in the extinct bovid *Myotragus balearicus*, from the Balearic Islands (Köhler & Moyà-Solà, 2009). In both cases, this peculiar growth pattern has been associated with potential periods of food scarcity, in the particular context of an insular environment. As previously pointed out, *Gargantuavis philoinos* inhabited the Ibero-Armorican Island in a period with a highly seasonal climate characterized by alternating dry and humid seasons (Smektala *et al.*, 2014). Therefore, it is possible that the type of growth of *Gargantuavis philoinos* was related to these environmental conditions.

So far, *Gargantuavis philoinos* has only been identified on the Ibero-Armorican Island (Figure 8.5); it apparently does not occur in the rich Late Cretaceous fossil vertebrate sites of other European regions (Romania, Hungary, etc.). Thus, it may have been the result of endemic evolution on this island; however, we know virtually nothing about the stages of this evolution. The evolution of large land birds in insular environments is a well-known phenomenon, including especially the case of moa in New Zealand and *Aepyornis* in Madagascar. However, these giant birds generally evolved in ecosystems that lacked land carnivores that might have threatened these flightless forms. This is not the case for the Ibero-Armorican Island, where *Gargantuavis* coexisted with various predators, especially theropod dinosaurs (Abelisauridae and Dromaeosauridae). Therefore, it remains to be explained how such a land bird could appear in such a type of ecosystem.

8.5.5. *Reproduction*

We know very little about the reproduction of *Gargantuavis*. It has been suggested that some of the numerous fossil eggs found in the continental Upper Cretaceous of France and Spain might have actually been laid by this giant bird (Buffetaut *et al.*, 1995; Buffetaut & Le Lœuff, 1998). Nevertheless, studies on the microstructure of the shells of large complete eggs, classically attributed to dinosaurs, do not reveal any typically avian features. Unlike dinosaurs, which in most cases buried their eggs, thus increasing the chances that entire eggs would be preserved, *Gargantuavis*, like present-day ratites, probably laid its eggs without any protection on the ground, so that the shells were not likely to be preserved entire, even when the eggs did not hatch.

8.6. Extinction

Gargantuavis philoinos is currently known from only a limited number of sites belonging to the Late Campanian/Early Maastrichtian interval. It has not been identified in Late Maastrichtian sites. In the region corresponding to the Ibero-Armorican Island, a significant change in the fauna took place during the Maastrichtian, involving in particular the mass arrival of hadrosaurid dinosaurs and the disappearance of the ornithopod *Rhabdodon* (Buffetaut *et al.*, 1997; Csiki-Sava *et al.*, 2015). The extinction of *Gargantuavis* may be linked to this faunal event. The possible reasons behind the disappearance of this giant bird, however, remain unknown, partly due to the lack of data about its lifestyle.

General Conclusions

After this overview about several groups of large land birds, which lived at very different periods (from the Late Cretaceous to the historical period) in various regions of the world, it is possible to find some points in common and regularities in the evolutionary history of these forms and to point out potentially significant differences.

An ancient and recurring phenomenon

A first significant point is the age of the phenomenon. Two simultaneous phenomena can be observed, namely the loss of the ability to fly and the evolution towards a large size, beginning in the Late Cretaceous with *Gargantuavis*, which was probably a primitive ornithurine.

Subsequently, the phenomenon took place again during the evolution of Neornithes, affecting both paleognaths and neognaths: Gastornithidae during the Paleogene, Dromornithidae from the Eocene to the Quaternary, Phorusrhacidae from the Eocene until possibly the Pleistocene, Brontornithidae during the Oligocene and the Miocene, Dinornithiformes from the Miocene until the historical period, and Aepyornithiformes in the Quaternary, possibly as late as the 17th Century. To these should be added the families of present-day ratites, which are represented by fossils dating back to the Tertiary, as well as forms of uncertain affinities, such as

Remiornis, and the large birds from the Paleogene of Africa, such as *Eremopezus*. Therefore, the evolution of such flightless and large forms is a frequent phenomenon in the evolution of birds.

Convergent evolution

The loss of the ability to fly and an increase in size can be observed in several groups of birds, which are not necessarily phylogenetically related. *Gargantuavis* occupies a special position due to its basal position among Ornithurines. Ratites remain somewhat united to the extent that they are paleognaths, but based on some recent molecular data, it is quite likely that the ability to fly was lost several times within this group, starting from volant ancestors. Gastornithidae, Dromornithidae and Brontornithidae are now generally regarded as Anseriformes, but their phylogenetic relationships remain obscure and there is no compelling evidence that they had a common flightless ancestor. Phorusrhacidae constitute a well-defined group within the order Cariamiformes. Nevertheless, it cannot be excluded that some of the smallest forms (Psilopterinae) were still able to fly.

It thus appears that the loss of the ability to fly combined with an increase in size is clearly a phenomenon of convergent evolution among birds. The environmental factors that led to these convergences are particularly worthy of considerations.

Locomotory and diet adaptations

The detailed analysis of the adaptations of the different groups considered reveals, besides shared features linked to flightlessness and large size, significant differences, especially in relation to their diet and type of locomotion.

As for diet, recent studies about functional anatomy and isotope geochemistry have shown that some birds long suspected of being carnivorous were in all likelihood herbivorous; this is the case for Gastornithidae, Dromornithidae and Brontornithidae. On the contrary, Phorusrhacidae have always been regarded as carnivores, and there is no reason to doubt this. The diet of *Gargantuavis* remains unknown.

Among the herbivorous forms, we can clearly distinguish ratites, both extant and extinct forms, with a small skull and fairly weak jaws from large Anseriformes (Gastornithidae, Dromornithidae and Brontornithidae) with a large skull and a strong beak. These differences certainly reveal different diets, as the forms with a strong beak could feed on tougher food.

In terms of locomotion, we can distinguish between cursorial and graviportal forms. While Gastornithidae, Dromornithidae and Brontornithidae were clearly graviportal, in other groups this varied from one genus to another. Thus, some moa were markedly graviportal, whereas others were much less so. Phorusrhacidae, a vast majority of which were cursorial, undoubtedly in relation to their carnivorous diet, nevertheless include some graviportal, robust and very large forms (*Paraphysornis* and *Kelenken*). It should be noted that all present-day large land birds (ratites) are cursorial, as all the graviportal forms have disappeared (their less rapid type of locomotion may have played a part in their extinction).

Geographical isolation and the structure of ecosystems

Some large flightless birds have had vast geographical distributions, spreading over several continents. This is the case for ostriches which, during the Neogene and until the Quaternary, were present in all of Africa and a large part of Eurasia. If we consider extinct groups, however, it appears that the evolution of large land birds is often associated with a certain degree of geographical isolation. The evolution of Aepyornithiformes in Madagascar and moa in New Zealand was clearly insular. In much older times, *Gargantuavis* seems to have been endemic to the Ibero-Armorican Island of the Late Cretaceous. Other groups evolved in clearly larger geographical areas which, however, remained geographically isolated for a long time. This is the case for the South American Phorusrhacidae during most of the Tertiary and the Australian Dromornithidae. Similarly, during the Paleocene, Gastornithidae evolved on the European continent, which was geographically isolated, even though the situation changed during the Eocene. This suggests that insularity, or at least a certain degree of geographical isolation, may have played a significant role in the evolution of these land birds. Nevertheless, isolation in itself is certainly not the only factor involved, as the structure of the ecosystems also played a very important part. The fact that there were no land mammals in New Zealand before the arrival of the Polynesians in the historical period and practically no predators certainly favored the evolutionary radiation of moa. In Madagascar, where there were very few large predators, the situation was very similar in relation to Aepyornithiformes. In other cases, the circumstances were quite different. *Gargantuavis* lived in environments inhabited by large carnivorous dinosaurs. The Gastornithidae developed in Europe during the Paleocene in an ecosystem inhabited by small mammals, where they represented the largest land animals; this situation changed in the Eocene when Europe was no longer isolated, but the Gastornithidae managed to survive among large mammals for a fairly long period of time. The Dromornithidae, in Australia, raise a slightly different problem as members of specific ecosystems that included large carnivorous mammals and reptiles. The situation was fairly similar in South America during the Tertiary, with the evolutionary radiation of large land birds in ecosystems that included numerous endemic mammals, whether herbivorous or carnivorous. Besides the herbivorous Brontornithidae, the Phorusrhacidae were unique insofar as they can be considered as the top predators of the food chain, a position they apparently did not manage to achieve durably in the other parts of the world where they are known to have lived, namely Africa and Europe.

Extinction

All the groups examined in this book are now extinct. Some of them disappeared in a far distant past, whereas others were clearly exterminated by humans only a few centuries ago, as the case of moa demonstrates more aptly than any others. Things are much less clear for those forms that became extinct before the appearance of humans, and the hypotheses put forward are often speculative, even if in some cases – such as that of the Phorusrhacidae – the competition with other groups of animals, made possible by the end of geographical isolation, may seem a plausible hypothesis. The fact that all the large birds that have survived until today are cursorial suggests that a graviportal type of locomotion may have represented a disadvantage in circumstances where changes in the environment (physical or living) exerted new selective pressures.

The large extinct land birds give us the possibility of studying, on the basis of fossil material that is often abundant, several significant evolutionary phenomena (among which the role of geographical isolation, of insularity and of the presence or absence of predators during the evolution of these birds, the acquisition of similar features in cases of convergent evolution, the diversity of adaptations, etc.). A very large number of points remain poorly understood in relation to these different aspects. We hope that this book will help identify with more precision problems that are still unresolved and provide insights into how they can be solved.

Bibliography

Acosta Hospitaleche C., Tambussi C.P., 2005, "Phorusrhacidae Psilopterinae (Aves) en la Formación Sarmiento de la localidad de Gran Hondonada (Eoceno superior), Patagonia, Argentina", Rev. Españ. Paleont., vol. 20, 121–132.

Acosta Hospitaleche C., Tambussi C.P., Reguero M., 2001, "Catálogo de tipos de aves fósiles del Museo de La Plata", Ser. Téc. Didáctica Rev. Mus. La Plata, vol. 41, 1–28.

Agnolin F., 2007, "*Brontornis burmeisteri* Moreno & Mercerat, un Anseriformes (Aves) gigante del Mioceno Medio de Patagonia, Argentina", Rev. Mus. Argent. Cienc. Nat., vol. 9, 15–25.

Agnolin F., 2009, "Sistemática y filogenia de las aves fororracoideas (Gruiformes: Cariamae)", Fundacion de Historia Natural Félix de Azara. Buenos Aires, Argentina.

Agnolin F., 2013, "La posición sistemática de *Hermosiornis* (Aves, Phororhacoidea) y sus implicancias filogenéticas", Rev. Mus. Argent. Cienc. Nat., vol. 15, 39–60.

Alexander R.McN., 1983, "Allometry of the leg bones of moas (Dinornithes) and other birds", J. Zool., vol. 200, 215–231.

Alexander R.McN., 1989, "Dynamics of dinosaurs and other extinct giants", Columbia University Press, New York.

Allentoft M.E., Rawlence N.J., 2012, "Moa's Ark or volant ghosts of Gondwana? Insights from nineteen years of ancient DNA research on the extinct moa (Aves: Dinornithiformes) of New Zealand", Ann. Anat.-Anat. Anz., vol. 194, 36–51.

Allentoft M.E., Heller R., Oskam C.L. *et al.*, 2014, "Extinct New Zealand megafauna were not in decline before human colonization", Proc. Natl. Acad. Sci., vol. 111, 4922–4927.

Alvarenga H.M.F., 1982, "Uma gigantesca ave fóssil do Cenozóico brasileiro: *Physornis brasiliensis* sp. n.", Acad. Bras. Ciênc. vol. 54, 697–712.

Alvarenga H.M.F., 1985, "Um novo Psilopteridae (Aves: Gruiformes) dos sedimentos terciarios de Itaborai, Rio de Janeiro, Brasil", Congresso Brasileiro de Paleontologia, NME-DNPM, Rio de Janeiro, pp. 17–20.

Alvarenga H.M.F., 1993, *Paraphysornis* novo gênero para *Physornis brasiliensis* Alvarenga, 1982 (Aves: Phorusrhacidae)", An. Acad. Bras. Ciênc. vol. 65, 403–406.

Alvarenga H.M.F., Höfling E., 2003, "Systematic revision of the Phorusrhacidae (Aves: Ralliformes)", Papéis Avulsos Zool., vol. 43, 55–91.

Alvarenga H.M.F., Jones W., Rinderknecht A., 2010, "The youngest record of phorusrhacid birds (Aves, Phorusrhacidae) from the late Pleistocene of Uruguay", N. Jb. Geol. Paläontol. Abh., vol. 256, 229–234.

Alvarenga H.M.F., Chiappe L., Bertelli S., 2011, "Phorusrhacids: the Terror Bird", in Dyke G., Kaiser G. (Eds.), Living Dinosaurs: The Evolutionary History of Modern Birds. Wiley, Chichester, pp. 187–208.

Amadon D., 1947, "An estimated weight of the largest known bird", The Condor, vol. 49, 159–164.

Ameghino F., 1887, "Enumeración systemática de las especies de mamíferos fósiles coleccionados por Carlos Ameghino en los terrenos eocenos de la Patagonia austral", Bol. Mus. Plata, vol. 1, 1–26.

Ameghino F., 1889, "Contribucion al conocimiento de los mamíferos fósiles de la República Argentina", Actas Acad. Nac. Cienc. Cordóba, vol. 6, 1–1027.

Ameghino F., 1891a, "Enumeración de las aves fósiles de la Republica Argentina", Rev. Argent. Hist. Nat., vol. 1, 441–453.

Ameghino F., 1891b, "Mamiferos y aves fósiles argentinas. - Especies nuevas, adiciones y correcciones", Rev. Argent. Hist. Nat., vol. 1, 240–259.

Ameghino F., 1895, "Sur les oiseaux fossiles de Patagonie", Bol. Inst. Geográfico Argent., vol. 15, 501–602.

Ameghino F., 1899, "Sinópsis geológico-paleontológico. Suplemento (adiciones y correcciones)", La Plata.

Ameghino F., 1901, "L'âge des formations sédimentaires de Patagonie", An. Soc. Cient. Argentina. Sociedad Científica Argentina, 65–91.

Anderson A., 1989, "Prodigious birds. Moas and moa-hunting in prehistoric New Zealand", Cambridge University Press, Cambridge.

Anderson J.F., Hall-Martin A., Russell D.A., 1985, "Long-bone circumference and weight in mammals, birds and dinosaurs", J. Zool. Lond., vol. 207, 53–61.

Anderson M.A., Fletcher H.O., 1934, "The Cuddie Springs bone bed", Aust. Mus. Mag., vol. 5, 152–158.

Andors A.V., 1988, "Giant groundbirds of North America (Aves, Diatrymidae)", Unpublished Ph.D. thesis Columbia University, New York, 577 p.

Andors A.V., 1992, "Reappraisal of the Eocene groundbird *Diatryma* (Aves: Anserimorphae)", in Papers in Avian Paleontology Honoring Pierce Brodkorb, Nat. Hist. Mus.Los Angeles Country Sc. Ser., Los Angeles, pp. 109–125.

Andors A.V., 1995, "*Diatryma* among the Dinosaurs", Nat. Hist., vol. 104, 68–71.

Andrews C.W., 1894, "Note on a new species of *Aepyornis* (*A. titan*)", Geol. Mag., 18–20.

Andrews C.W., 1897, "On a nearly complete skeleton of *Aepyornis hildebrandti*", Geol. Mag., 241–250.

Andrews C.W., 1899, "On the extinct birds of Patagonia – The skull and skeleton of *Phorusrhacos inflatus* Ameghino", Trans. Zool. Soc. Lond., vol. 15, 55–86.

Andrews C.W., 1904, "On the pelvis and hind-limb of *Mullerornis betsilei*", Proc. Zool. Soc. Lond., 163–171.

Andrews C.W., 1917, "A gigantic Eocene bird", Geol. Mag., vol. 4, 469–471.

Angst D., 2014, "Successeurs des dinosaures ? Paléobiologie et paléoécologie des oiseaux géants terrestres du Crétacé supérieur et du Paléocène", Thesis, UCBL, Lyon.

Angst D., Buffetaut E., 2013, "The first mandible of *Gastornis* HÉBERT, 1855 (Aves, Gastornithidae) from the Thanetian (Paleocene) of Mont-de-Berru (France)", Rev. Paléobiol. Geneva, vol. 32, 423–432.

Angst D., Buffetaut E., Lécuyer C. *et al.*, 2013, "Terror Birds" (Phorusrhacidae) from the Eocene of Europe Imply Trans-Tethys Dispersal", PloS One, vol. 8, 1–8.

Angst D., Chinsamy A., 2016, "Ecological implications of the revised locomotion of the South American giant birds: *Paraphysornis* and *Brontornis*", in 9th International Meeting of the Society of Avian Paleontology and Evolution, Diamante, Abstracts.

Angst D., Chinsamy A., in press, "Ecological implications of the revised locomotory habits of the giant extinct South American birds (Phorusrhacidae and Brontornithidae)", Rev. Mus. Argent. Cienc. Nat.

Angst D., Lécuyer C., Amiot R. *et al.*, 2014a, "Isotopic and anatomical evidence of an herbivorous diet in the Early Tertiary giant bird *Gastornis*. Implications for the structure of Paleocene terrestrial ecosystems", Naturwissenschaften, vol. 101, 313–322.

Angst D., Buffetaut E., Lécuyer C. *et al.*, 2014b, "Fossil avian eggs from the Palaeogene of southern France: new size estimates and a possible taxonomic identification of the egg-layer", Geol. Mag., vol. 152, 1–10.

Angst D., Amiot R., Buffetaut, E. *et al.*, 2015, "Diet and climatic context of giant birds inferred from $\delta^{13}C_c$ and $\delta^{18}O_c$ values of Late Palaeocene and Early Eocene eggshells from southern France", Palaeogeogr. Palaeoclimatol. Palaeoecol., vol. 435, 210–221.

Angst D., Buffetaut E., Lécuyer C. *et al.*, 2016, "A new method for estimating locomotion type in large ground birds", Palaeontology, vol. 59, 217–223.

Angst D., Buffetaut E., Corral J.C. *et al.*, 2017, "First record of the Late Cretaceous giant bird *Gargantuavis philoinos* from the Iberian Peninsula", Annls. Paléont., vol. 103, 135–139.

Angst D., Chinsamy A., Steel L., Hume J.P., 2017, "Bone histology sheds new light on the ecology of the dodo (*Raphus cucullatus*, Aves, Columbiformes)", Scientific Reports, vol. 7, 1–10.

Ar A., Paganelli C.V., Reeves R.B. *et al.*, 1974, "The avian egg: water vapor conductance, shell thickness, and functional pore area", The Condor, vol. 76, 153–158.

Archer M., Hand S.J., Godthelp H., 1994, "Riversleigh", in Riversleigh: The story of animals in ancient rainforest of inland Australia, AH & AW Re, Sydney.

Archer M., Hand S.J., Godthelp H. *et al.*, 1997, "Correlation of the Cenozoic sediments of the Riversleigh World Heritage Fossil Property, Queensland, Australia", in Aguilar J.-P., Legendre S., Michaux J. (Eds.), BiochroM '97, Mém. Trav. Inst. Ecole Prat. Hautes Etudes, Montpellier, pp. 131–152.

Archey G., 1941, "The Moa: a study of the Dinonithiformes", Bull. Auckl. Inst. Mus., vol. 1, 1–135.

Atkinson I.A., Greenwood R.M., 1989, "Relationships between moas and plants", N. Z. J. Ecol., vol. 12, 67–96.

Attard M.R.G., Wilson L.A.B., Worthy T.H. *et al.*, 2016, "Moa diet fits the bill: virtual reconstruction incorporating mummified remains and prediction of biomechanical performance in avian giants", Proc. Roy. Soc. B, vol. 283, 2015–2043.

Baker A.J., Haddrath O., McPherson J.D. *et al.*, 2014, "Genomic support for a Moa–Tinamou clade and adaptive morphological convergence in flightless ratites", Mol. Biol. Evol., vol. 31, 1686–1696.

Balanoff A.M., Rowe T., 2007, "Osteological description of an embryonic skeleton of the extinct elephant bird, *Aepyornis* (Palaeognathae: Ratitae)", J. Vertebr. Paleontol., vol. 27, 1–53.

Baskin J.A., 1995, "The giant flightless bird *Titanis walleri* (Aves: Phorusrhacidae) from the Pleistocene coastal plain of south Texas", J. Vertebr. Paleontol., vol. 15, 842–844.

Battistini R., Vérin P., 1967, "Ecologic changes in protohistoric Madagascar", in Martin P.S., Wright Jr. H.E. (Eds.), Pleistocene Extinctions. Yale University Press, New Haven, pp. 407–424.

Battistini R., Vérin P., 1972, "Man and the environment in Madagascar", in Battistini R., Richard-Vindard G. (Eds.), Biogeography and Ecology in Madagascar. W. Junk B.V., Springer Science & Business Media, pp. 311–337.

Battistini R., Vérin P., Raison R., 1963, "Le site archéologique de Talaky: cadre géographique", Ann. Malgaches, Faculté des Lettres et Sciences Humaines, vol. 1, 112–153.

Baumel J.J., King A.S., Breazile J.E. *et al.*, 1993, "Handbook of Avian Anatomy: Nomina Anatomica Avium", Second Edition, Publication of the Nuttall Ornithological Club, Cambridge, Massachusetts, 779 p.

Berentson Q., 2012, "Moa. The life and death of New Zealand's legendary bird", Craig Potton Publishing., Nelson.

Berg D.E., 1965, "Nachweis des Riesenlaufenvogels *Diatryma* im Eozän von Messel bei Darmstadt/Hessen", Notizbl. Hess. Landesamtes Bodenforsch., vol. 93, 68–72.

Berger R., Ducote K., Robinson K. *et al.*, 1975, "Radiocarbon date for the largest extinct bird", Nature, vol. 258, 709.

Bernard A., Daux V., Lécuyer C. *et al.*, 2009, "Pleistocene seasonal temperature variations recorded in the $\delta^{18}O$ of *Bison priscus* teeth", Earth Planet. Sci. Lett., vol. 283, 133–143.

Bertelli S., Chiappe L.M., Tambussi C.P., 2007, "A new Phorusrhacid (Aves: Cariamae) from the middle Miocene of Patagonia, Argentina", J. Vertebr. Paleontol., vol. 27, 409–419.

Bertram B.C.R., 1980, "Vigilance and group size in ostriches", Anim. Behav., vol. 28, 278–286.

Bertram B.C.R., 1992, "The ostrich communal nesting system", Princeton University Press, Princeton.

Best L.B., 1995, "Grit-use behavior in birds: a review of research to develop safer granular pesticides", in National Wildlife Research Center Repellents Conference. USDA National Wildlife Research Center Symposia, Lincoln, pp. 333–341.

Bianconi G.-G., 1865, "Recherches sur les os de l'*Aepyornis maximus*", C. R. Acad. Sci., 179–180.

Bianconi M.J.J., 1870, "Recherches sur les affinités naturelles de l'*Aepyornis*", Ann. Sci. Nat., 3–5.

Black K.H., Archer M., Hand S.J. *et al.*, 2012, "The rise of Australian marsupials: a synopsis of biostratigraphic, phylogenetic, palaeoecologic and palaeobiogeographic understanding", in Talent J.A. (Ed.), Earth and life: Global biodiversity, extinction intervals and biogeographic perturbations through time. Springer, Dordrecht, pp. 983–1078.

Blackburn G., Allison G.B., Leaney F.W.J., 1982, "Further evidence on the age of the tuff at Mount Gambier, South Australia", R. Soc. S. Aust. Trans., vol. 106, 163–167.

Blanco R.E., Jones W.J., 2005, "Terror birds on the run: a mechanical model to estimate its maximum running speed", Proc. Roy. Soc. Lond., vol. 272, 1769–1773.

Bledsoe A.H., 1988, "A phylogenetic analysis of postcranial skeletal characters of the ratite birds", Ann. Carnegie Mus., vol. 57, 73–90.

Bonaparte C.L., 1849, "Conspectus systematis ornithologiae. Editio altera reformata additis synonymis Grayanis", M. Westerman & fils. Amsterdam.

Bonaparte C.L., 1853, "Classification ornithologique par séries de Bonaparte", C. R. Acad. Sci., vol. 37, 641–647.

Bonaparte C.L., 1856, "Conspectus ineptorum et struthiorum", C. R. Acad. Sci., vol. 43, 840–841.

Bond W.J., Silander J.A., 2007, "Springs and wire plants: anachronistic defences against Madagascar's extinct elephant birds", Proc. R. Soc. Lond. B Biol. Sci., vol. 274, 1985–1992.

Bourdon E., Cracraft J., 2011, "*Gastornis* is a terror bird: New insights into the evolution of the Cariamae (Aves, Neornithes)", 71st Ann. Meet. Soc. Vert. Paleont. Progr. Abstr., p. 75

Bourdon E., Castanet J., De Ricqlès A. *et al.*, 2009, "Bone growth marks reveal protracted growth in New Zealand kiwi (Aves, Apterygidae)", Biol. Lett., vol. 5, 639–642.

Bourdon E., Mourer-Chauviré C., Laurent Y., 2014, "Early Eocene birds from La Borie, southern France", Acta Palaeontol. Pol., vol. 61, 175–190.

Bravard A., 1860, "Catalogue des espèces d'animaux fossiles recueillies dans l'Amérique du Sud par Auguste Bravard, 1860", Document autolithographié.

Brodkorb P., 1963, "A giant flightless bird from the Pleistocene of Florida", The Auk, vol. 80, 111–115.

Brodkorb P., 1967, "Catalogue of fossil birds: part 3 (Ralliformes, Ichthyornithiformes, Charadriiformes)", Bull. Fla. State Mus., vol. 11, 99–220.

Buffetaut E., 1997a, "New remains of the giant bird *Gastornis* from the Upper Paleocene of the eastern Paris Basin and the relationships between *Gastornis* and *Diatryma*", N. Jb. Geol. Paläont. Mh., vol. 3, 179–190.

Buffetaut E., 1997b, "L'oiseau géant *Gastornis* : interprétation, reconstitution et vulgarisation de fossiles inhabituels dans la France du XIXe siècle", Bull. Soc. Géol. Fr., vol. 168, 805–811.

Buffetaut E., 2000, "Are *Gastornis* and *Diatryma* congeneric?", Vertebr. Palasiat., vol. 38, 3.

Buffetaut E., 2004, "Footprints of giant birds from the Upper Eocene of the Paris Basin: an ichnological enigma", Ichnos, vol. 11, 357–362.

Buffetaut E., 2008, "First evidence of the giant bird *Gastornis* from southern Europe: a tibiotarsus from the Lower Eocene of Saint-Papoul (Aude, southern France)", Oryctos, vol. 7, 75–82.

Buffetaut E., 2013a, "From Sinbad the sailor to H. G. Wells: *Aepyornis* and the Rûkh bird", in Talairach-Vielmas L., Bouchet M. (Eds.), Lost and Found: In Search of Extinct Species. Muséum d'Histoire Naturelle de Toulouse, pp. 159–166.

Buffetaut E., 2013b, "The giant bird *Gastornis* in Asia: A revision of *Zhongyuanus xichuanensis* Hou, 1980, from the Early Eocene of China", Paleontol. Zh. vol. 47, 1302–1307.

Buffetaut E., 2013c, "Phororhacoidea or Phorusrhacoidea? A note on the nomenclature of the "terror birds"", Ann. Paléont., vol. 99, 157–161.

Buffetaut E., 2013d, "Who discovered the Phorusrhacidae? An episode in the history of avian palaeontology", in Göhlich U.B., Kroh A. (Eds.), Paleornithological Research 2013, Proceedings 8th International Meeting of the Society of Avian Paleontology and Evolution, Wien, Naturhistorisches Museum, pp. 123–133.

Buffetaut E., 2014, "Tertiary ground birds from Patagonia (Argentina) in the Tournouër collection of the Muséum National d'Histoire Naturelle, Paris", Bull. Soc. Geol. Fr., vol. 185, 207–214.

Buffetaut E., 2016a, "From Charles Darwin's comments to the first mention of South American giant fossil birds: Auguste Bravard's catalogue of fossil species from Argentina (1860) and its significance", Bull. Soc. Géol. France, vol. 187, 41–53.

Buffetaut E., 2016b, "A reassessment of the giant birds *Liornis floweri* Ameghino, 1895 and *Callornis giganteus* Ameghino, 1895, from the Santacrucian (late Early Miocene) of Argentina", Palaeovertebrata, vol. 40, e3.

Buffetaut E., in press, A brontornithid from the Deseadan (Oligocene) of Bolivia. Rev Mus Argent. Cienc Nat.

Buffetaut E., Angst D., 2013, "New evidence of a giant bird from the Late Cretaceous of France", Geol. Mag., vol. 150, 173–176.

Buffetaut E., Angst D., 2014a, "Stratigraphic distribution of large flightless birds in the Palaeogene of Europe and its palaeobiological and palaeogeographical implications", Earth-Sci. Rev., vol. 138, 394–408.

Buffetaut E., Angst D., 2014b, "Stratigraphic distribution of large flightless birds in the Paleogene of Europe", in Proceedings of the First International Congress on Stratigraphy, pp. 1005–1008.

Buffetaut E., Angst D., 2016a, "The giant flightless bird *Gargantuavis philoinos* from the Late Cretaceous of southwestern Europe: a review", in Khosla A., Lucas S.G. (Eds.), Cretaceous Period: Biotic Diversity and Biogeography, New Mexico Mus. Nat. Hist. Sc. Bull., pp. 45–50.

Buffetaut E., Angst D., 2016b, "Pelvic elements of the giant bird *Gargantuavis* from the Upper Cretaceous of Cruzy (southern France), with remarks on pneumatisation", Cretac. Res., vol. 66, 171–176.

Buffetaut E., Angst D., Mechin P. *et al.*, 2015, "New remains of the giant bird *Gargantuavis philoinos* from the Late Cretaceous of Provence (south-eastern France)", Palaeovertebrata, vol. 39, 1–6.

Buffetaut E., Le Lœuff J., 1998, "A new giant ground bird from the Upper Cretaceous of southern France", J. Geol. Soc. Lond., vol. 155, 1–4.

Buffetaut E., Le Lœuff J., 2010, "*Gargantuavis philoinos*: giant bird or giant pterosaur?" Ann. Paléont., vol. 96, 135–141.

Buffetaut E., Le Lœuff J., Mechin P. *et al.*, 1995, "A large French Cretaceous bird", Nature, vol. 377, 110.

Buffetaut E., Le Lœuff J., Cavin L. *et al.*, 1997, "Late Cretaceous non-marine vertebrates from southern France: A review of recent finds", Geobios Mém. Spéc., vol. 20, 95–102.

Buick T.L., 1931, "The mystery of the Moa", Thomas Avery & Sons., New Plymouth.

Bunce M., Worthy T.H., Ford T. *et al.*, 2003, "Extreme reversed sexual size dimorphism in the extinct New Zealand moa *Dinornis*", Nature, vol. 425, 172–175.

Bunce M., Worthy T.H., Phillips M.J. *et al.*, 2009, "The evolutionary history of the extinct ratite moa and New Zealand Neogene paleogeography", Proc. Natl. Acad. Sci., vol. 106, 20646–20651.

Burckhardt R., 1893, "Über *Aepyornis* (*A. hildebrandti*)", Palaeont. Abh., vol. 6, 127–145.

Burney D., James H., Grady F. *et al.*, 1997, "Environmental change, extinction and human activity: evidence from caves in NW Madagascar", J. Biogeogr., vol. 24, 755–767.

Burney D.A., 1993, "Late Holocene environmental changes in arid southwestern Madagascar", Quat. Res., vol. 40, 98–106.

Burney D.A., 1997, "Theories and facts regarding Holocene environmental change before and after human colonization", in Goodman S.M., Patterson B.D. (Eds.), Natural Change and Human Impact in Madagascar. Washington D.C., pp. 75–89.

Burney D.A., 1999, "Rates, patterns, and processes of landscape transformation and extinction in Madagascar", in MacPhee, R.D.E. (Ed.), Extinction in Near Time, Kluwer, New York, pp. 145–164.

Burney D.A., Burney L.P., Godfrey L.R. *et al.*, 2004, "A chronology for late prehistoric Madagascar", J. Hum. Evol., vol. 47, 25–63.

Burrows C.J., 1980, "Diet of New Zealand Dinornithiformes", Naturwissenschaften, vol. 67, 151–153.

Butler W.H., 1969, "Remains of *Sarcophilus* the "Tasmanian" Devil (Marsupialia, Dasyuridae) from coastal dunes South of Scott River, Western Australia", West. Aust. Nat., vol. 11, 87–88.

Cabrera A., 1939, "Sobre vertebrados fósiles del Plioceno de Adolfo Alsina", Rev. Mus. Plata, vol. 2, 3–35.

Cailleux A., 1969a, "Observations sur des gastrolithes d'oiseaux holocènes et tertiaires", C. R. Soc. géol. Fr., vol. 73–75.

Cailleux A., 1969b, "Pierres de gésier au microscope électronique à balayage", Sci. Prog. Nat., 224–226.

Calder W.A., 1984, "Size, function, and life history", Dover, New York.

Campbell Jr K.E., Marcus L., 1992, "The relationship of hindlimb bone dimensions to body weight in birds", Nat. Hist. Mus. Los Angel. Cty. Sci. Ser., vol. 36, 395–412.

Campbell Jr K.E., Tonni E., 1983, "Size and locomotion in Teratorns (Aves: Teratornithidae)", The Auk, vol. 100, 390–403.

Campbell B., Lack E., 1985, "A dictionary of birds", Buteo Books, Vermilion.

Capellini G., 1889, "Sul primo uovo di *Aepyornis maximus* arrivato in Italia nel 1888", Mem. R. Accad. Sci. Bologna., vol. 4, 23–42.

Carrano M.T., 1999, "What, if anything, is a cursor? Categories versus continua for determining locomotor habit in mammals and dinosaurs", J. Zool., vol. 247, 29–42.

Castanet J., Francillon-Vieillot H., Meunier F.J. *et al.*, 1993, "Bone and individual aging", in Bone growth (ed. Hall B.K.), Boca Raton, FL: CRC Press, pp. 245–283.

Cenizo M.M., Tambussi C.P., Montalvo C.I., 2012, "Late Miocene continental birds from the Cerro Azul Formation in the Pampean region (central-southern Argentina)", Alcheringa, vol. 36, 47–68.

Cerling T.E., Harris J.M., 1999, "Carbon isotope fractionation between diet and bioapatite in ungulate mammals and implications for ecological and paleoecological studies", Oecologia, vol. 120, 347–363.

Cerling T.E., Harris J.M., MacFadden B.J. *et al.*, 1997, "Global vegetation change through the Miocene/Pliocene boundary", Nature, vol. 389, 153–158.

Chan N.R., 2014, "Does size variation in *Genyornis newtoni* (Aves, Dromornithidae) encompass eggshell safety limits?", J. Vertebr. Paleontol., vol. 34, 976–979.

Chandler R.M., 1994, "The wing of *Titanis walleri* (Aves: Phorusrhacidae) from the Late Blancan of Florida", Bull. Fla. Mus. Nat. Hist. Biol. Sci., vol. 36, 175–180.

Chinsamy A., 1995, "Histological perspectives on growth in the birds *Struthio camelus* and *Sagittarius serpentarius*". Cour. Forschungsinstitut Senckenberg, vol. 181, 317–323.

Chinsamy A., Chiappe L.M., Marugán-Lobón J. *et al.*, 2013, "Gender identification of the Mesozoic bird *Confuciusornis sanctus*", Nat. Commun., vol. 4, 1–5.

Chinsamy A., Buffetaut E., Canoville A. *et al.*, 2014, "Insight into the growth dynamics and systematic affinities of the Late Cretaceous *Gargantuavis* from bone microstructure", Naturwissenschaften, vol. 101, 447–452.

Clarke S.J., Miller G.H., Fogel M.L. *et al.*, 2006, "The amino acid and stable isotope biogeochemistry of elephant bird (*Aepyornis*) eggshells from southern Madagascar", Quat. Sci. Rev., vol. 25, 2343–2356.

Coddington C.L., Cockburn A., 1995, "The mating system of free-living emus", Aust. J. Zool., vol. 43, 365–372.

Coombs Jr W.P., 1978, "Theoretical aspects of cursorial adaptations in dinosaurs", Q. Rev. Biol., vol. 53, 393–418.

Cooper A., Lalueza-Fox C., Anderson S. *et al.*, 2001, "Complete mitochondrial genome sequences of two extinct moas clarify ratite evolution", Nature, vol. 409, 704–707.

Cope E.D., 1876, "On a gigantic bird from the Eocene of New Mexico", Proc. Acad. Nat. Sci. Phila., vol. 28, 10–12.

Cope E.D., 1877, "Report upon the extinct vertebrata obtained in New Mexico by parties of the expedition of 1874", in Geographical Surveys West of the One Hundredth Meridian, Paleontology. Government Printing Office, Washington, D.C., p. 370.

Coplen T.B., Kendall C., Hopple J., 1983, "Comparison of stable isotope reference samples", Nature, vol. 302, 236–238.

Corral J.-C., Pueyo E.L., Berreteaga A. *et al.*, 2016, "Magnetostratigraphy and lithostratigraphy of the Laño vertebrate-site: implications in the uppermost Cretaceous chronostratigraphy of the Basque-Cantabrian Region", Cretac. Res., vol. 57, 473–489.

Corroy G., Touraine F., 1961, "Données nouvelles sur la stratigraphie du Grand-Cabri et du Cengle (Bouches-du-Rhône)", Bull. Soc. géol. Fr., vol. 7, 77–81.

Covacevich V., Rich P., 1982, "New bird ichnites from Fildes Peninsula, King George Island, West Antarctica", in Craddock, C. (Ed.), Antarctic Geoscience. University of Wisconsin Press, Madison, pp. 245–254.

Cracraft J., 1974, "Phylogeny and evolution of the ratite birds", Ibis, vol. 116, 494–521.

Cracraft J., 1976, "The species of moas (Aves: Dinornithidae)", Smithson. Contrib. Paleobiol., vol. 27, 189–205.

Crome F.H.J., 1975, "Some observations on the biology of the cassowary in northern Queensland", Emu, vol. 76, 8–14.

Crowley B.E., 2010, "A refined chronology of prehistoric Madagascar and the demise of the megafauna", Quat. Sci. Rev., vol. 29, 2597–2601.

Csiki-Sava Z., Buffetaut E., Ösi A. *et al.*, 2015, "Island life in the Cretaceous-faunal composition, biogeography, evolution, and extinction of land-living vertebrates on the Late Cretaceous European archipelago", ZooKeys, vol. 469, 1–161.

Davies S.J.J.F., 2002, "Bird families of the world. Ratites and Tinamous: Tinamidae, Rheidae, Dromaiidae, Casuariidae, Apterygidae, Struthionidae", Oxford University Press, Oxford.

De Geyter, G., De Man, E., Herman, J. *et al.*, 2006, "Disused Paleogene regional stages from Belgium: Montian, Heersian, Landenian, Paniselian, Bruxellian, Laekenian, Ledian, Wemmelian and Tongrian", Geol. Belg., vol. 9, 203–213.

Decary R., 1937, "La légende du Rokh et l'*Aepyornis*", Bull. Acad. Malgache Nouv. Sér., vol. 20, 107–113.

Deeming D.C., 1999, The ostrich: biology, production and health. CABI, Oxon.

Degrange F.J., Noriega J.I., Areta J.I., 2012, "Diversity and paleobiology of the Santacrucian birds", in Vizcaíno S.F., Kay R.F., Bargo M.S. (Eds.), Early Miocene Paleobiology in Patagonia: High-Latitude Paleocommunities of the Santa Cruz Formation. Cambridge University Press, Cambridge, pp. 138–155.

Degrange F.J., Tambussi C.P., 2011, "Re-examination of *Psilopterus lemoinei* (Aves, Phorusrhacidae), a late early Miocene little Terror bird from Patagonia (Argentina)", J. Vertebr. Paleontol., vol. 31, 1080–1092.

Degrange F.J., Tambussi C.P., Moreno K. *et al.*, 2010, "Mechanical analysis of feeding behavior in the extinct "Terror Bird" *Andalgalornis steulleti* (Gruiformes: Phorusrhacidae)", PLoSOne, vol. 5, e11856. doi:10.1371/journal.pone.0011856.

Degrange F.J., Tambussi C.P., Taglioretti M.L. *et al.*, 2015, "A new Mesembriornithinae (Aves, Phorusrhacidae) provides new insights into the phylogeny and sensory capabilities of terror birds", J. Vertebr. Paleontol., vol. 35, e912656.

Deschamps H., 1961, "Histoire de Madagascar", Berger-Levrault.

Desnoyers J., 1859a, "Note sur des empreintes de pas d'animaux dans le gypse des environs de Paris, particulièrement de la vallée de Montmorency", C. R. Acad. Sci. Paris, vol. 49, 67–73.

Desnoyers J., 1859b, "Sur des empreintes de pas d'animaux dans le Gypse des environs de Paris, et particulièrement de la vallée de Montmorency", Bull. Soc. géol. Fr., vol. 16, 936–944.

DeVis C.W., 1884, "The moa (*Dinornis*) in Australia", Proc. R. Soc. Qld., vol. 1, 23–28.

DeVis C.W., 1891, "The moa in Australia", N. Z. J. Sci., vol. 1, 97–101.

Dickison M.J., 2007, "Allometry of giant flightless birds", Unpublished PhD thesis, Duke University, Durham NC, USA.

Diederle J.M., Noriega J.I., 2013, "Aves del Mioceno de la Provincia de Entre Rios, Argentina", in Brandoni D., Noriega J.I. (Eds.), El Neógeno de La Mesopotamia Argentina, Asoc. Paleont. Argent., Pub. Espec. pp. 97–108.

Diefendorf A.F., Mueller K.E., Wing S.L. *et al.*, 2010, "Global patterns in leaf ^{13}C discrimination and implications for studies of past and future climate", Proc. Natl. Acad. Sci., vol. 107, 5738–5743.

Dolgopol de Saez M., 1927, "Las aves corredoras fósiles del Santacrucense", Soc. Cient. Arg., vol. 103, 145–160.

Dollo M.L., 1883, "Note sur la présence du *Gastornis edwardsii*, Lemoine dans l'assise inférieure de l'étage Landénien, à Mesvin, près de Mons", Bull. Mus. R. Hist. Nat. Belg., vol. 2, 297–308.

Dollo M.L., 1909, "The fossil vertebrates of Belgium", Ann. N. Y. Acad. Sci., vol. 19, 99–119.

Donaire M., Lopez-Martinez N., 2009, "Porosity of Late Paleocene *Ornitholithus* eggshells (Tremp Fm, south-central Pyrenees, Spain): Paleoclimatic implications", Palaeogeogr. Palaeoclimatol. Palaeoecol., vol. 279, 147–159.

Dorst J., 1982, "The life of birds", Weidenfeld & Nicolon. London.

Dughi R., Sirugue F., 1959, "Sur les fragments de coquilles d'œufs fossiles de l'Éocène de Basse-Provence", C. R. Hebd. Séances Acad. Sci., vol. 249, 959–961.

Dughi R., Sirugue F., 1962, "Distribution verticale des œufs d'oiseaux fossiles de l'Éocène de Basse-Provence", Bull. Soc. géol. Fr., vol. 4.

Dughi R., Sirugue F., 1968, "Marnes à œufs d'oiseaux du Paléocène de Basse-Provence", Bull. Soc. géol. Fr., vol. 10, 542–548.

Dulhunty J.A., 1971, "Potassium-argon basalt dates and their significance in the Ilford-Mudgee-Gulgong region", J. Proc. R. Soc. New South Wales, vol. 104, 39–44.

Duméril C., 1855, "Sur une configuration caractéristique du tibia des oiseaux", C. R. Acad. Sci., vol. 40, 619–620.

Duprat M., Lecomte G., Buffetaut E. *et al.*, 2010, "Some exceptional MP6 to MP16 estuarine channels at Paleogene stage boundaries in the Paris Basin, E of the Pays de Bray anticline in Marne (Louvois MP6 to Monthelon MP10b) and Aisne (Condé-en-Brie MP9 to Chéry-Chartreuve/Rocourt-St-Martin MP15b)", 4th French Stratigraphy Congress, Paris, pp. 86–88.

Dyke G.J., Kaiser G.W., 2010, "Cracking a developmental constraint: egg size and bird evolution", Rec. Aust. Mus., vol. 62, 207–216.

Erben H.K., Hoefs J., Wedepohl K.H., 1979, "Paleobiological and isotopic studies of eggshells from a declining dinosaur species", Paleobiology, vol. 5, 380–414.

Etheridge R., 1889, "On further evidence of a large extinct struthious bird (*Dromornis* Owen) from the post Tertiary deposits of Queensland", Rec. Geol. Surv. New South Wales, vol. 89, 126–136.

Fabre-Taxy S., Touraine F., 1960, "Gisements d'œufs d'oiseaux de très grande taille dans l'Éocène de Provence", C. R. Acad. Sci., vol. 250, 3870–3871.

Feduccia A., 2014, "Avian extinction at the end of the Cretaceous: Assessing the magnitude and subsequent explosive radiation", Cret. Res., vol. 50, 1–15.

Field J.H., Boles W.E., 1998, "*Genyornis newtoni* and *Dromaius novaehollandiae* at 30,000 B.P. in central northern New South Wales", Alcheringa, vol. 22, 177–188.

Fischer K., 1962, "Der Riesenlaufvogel *Diatryma* aus der eozänen Braunkohle des Geiseltales", Hallesches Jahrb. Mitteldtsch. Erdgesch., vol. 4, 26–33.

Fischer K., 1978, "Neue Reste des Riesenlaufvogels *Diatryma* aus dem Eozän des Geiseltales bei Halle (DDR)", Mitt. Zool. Mus. Berlin, vol. 2, 133–144.

Flacourt E. de, 1661, "Histoire de la grande isle Madagascar", Gervais Clouzier, Paris

Folch A., 1992, "Struthioniformes", in Del Hoyo J., Elliott A., Sargatal J. (Eds.), Handbook of the birds of the world. Lynx Edicions, Barcelona, vol. 1, 76–110.

Fondevilla,V., Dinarès-Turell J., Vila B. *et al.*, 2016, "Magnetostratigraphy of the Maastrichtian continental record in the Upper Aude Valley (northern Pyrenees, France): Placing age constraints on the succession of dinosaur-bearing sites", Cretac. Res., vol. 57, 457–472.

Franzen J.L., 2005, "The implications of the numerical dating of the Messel fossil deposit (Eocene, Germany) for mammalian biochronology", Ann. Paléont., vol. 329–335.

Fürbringer M., 1888, "Untersuchungen zur Morphologie und Systematik der Vögel, zugleich ein Beitrag zur Anatomie der Stütz und Bewegungsorgane", Tj. Van Holkema, Amsterdam, pp. 1–1751.

Gadow H., 1893, "Vögel, part II: Systematischer Theil", Dr H. G. Bronn's Classen und Ordnungen des Thierreichs. Leipzig.

Gaillard C., 1936, "Un oiseau géant dans les dépôts éocènes du Mont-d'Or lyonnais", C. R. Acad. Sci., vol. 202, 965–967.

Gaillard C., 1937, "Un oiseau géant dans les dépôts éocènes du Mont-d'Or Lyonnais", Ann. Soc. Linn. Lyon, vol. 80, 111–126.

Geoffroy Saint-Hilaire I., 1851, "Des ossements et des œufs trouvés à Madagascar, dans des alluvions modernes, et provenant d'un oiseau gigantesque", C. R. Hebd. Sc. Acad. Sci., vol. 32, 101–107.

Gill B.J., 2006, "A catalogue of moa eggs (Aves: Dinornithiformes)", Rec. Auckl. Mus., vol. 43, 55–80.

Gill B.J., 2007, "Eggshell characteristics of moa eggs (Aves: Dinornithiformes)", J. R. Soc. N. Z., vol. 37, 139–150.

Gill E., 1972, "Eruption date of Tower Hill Volcano, Western Victoria, Australia", Vic. Nat., vol. 89, 188–192.

Gillespie R., 1978, "Lancefield swamp and the extinction of the Australian megafauna", Science, vol. 200, 1044–1048.

Glauert L., 1910, "The Mammoth Cave", Rec. West. Aust. Mus., vol. 1, 11–36.

Glauert L., 1914, "The Mammoth Cave", Rec. West. Aust. Mus., vol. 1, 244–252.

Glauert L., 1948, "The cave fossils of the south-west", West. Aust. Nat., vol. 1, 101–104.

Gommery D., Sénégas F., Valentin F. *et al*, 2011, "Madagascar : Premiers habitants et biodiversité passée", Archéologia, vol. 40–49.

Goodman S.M., 1999, "Holocene bird subfossils from the sites of Ampasambazimba, Antsirabe and Ampoza, Madagascar: changes in the avifauna of south central Madagascar over the past few millennia", in Adams N.J., Slotow R.H. (Eds.), Proceedings of the 22nd International Ornithological Congress. Johannesburg, pp. 3071–3083.

Goodman S.M., Jungers W.L., 2013, "Extinct Madagascar: Picturing the island's past", University of Chicago Press, Chicago.

Gould G.C., Quitmyer I.R., 2005, "*Titanis walleri*: Bones of contention", Bull. Fla. Mus. Nat. Hist., vol. 45, 201–229.

Grande L., 2013, "The lost world of fossil lake: snapshots from deep time", University of Chicago Press, London & Chicago.

Grandidier A., 1867a, "Observations sur le gisement des œufs de l'*Aepyornis*", C. R. Acad. Sci., vol. 476–478.

Grandidier A., 1867b, "Lettre rectifiant la position géographique des principales rivières de la côte sud-est et annonçant la découverte d'ossements fossiles d'un nouvel *Aepyornis* et d'un Hippopotame", Bull. Soc. Geogr. Fr., vol. 508–510.

Grandidier A., 1868, "Observation of the egg-beds of *Aepyornis*", Ibis, vol. 4, 65–68.

Greenwood R.M., Atkinson I.A., 1977, "Evolution of divaricating plants in New-Zealand in relation to moa browsing", Proc. N. Z. Ecol. Soc., vol. 24, 21–33.

Gregory W.K., 1912, "Notes on the principles of quadrupedal locomotion and on the mechanism of the limbs in hoofed animals", Ann. N. Y. Acad. Sci., vol. 22, 267–294.

Grellet-Tinner G., Spooner N.A., Worthy T.H., 2016, "Is the "*Genyornis*" egg of a mihirung or another extinct bird from the Australian dreamtime?", Quat. Sci. Rev., vol. 133, 147–164.

Gunn R.G., Douglas L.C., Whear R.L., 2011, "What bird is that? Identifying a probable painting of *Genyornis newtoni* in Western Arnhem Land", Aust. Archaeol., 1–12.

Haast J., 1874, "Anniversary address of the President", Trans. Proc. N. Z. Inst., vol. 6, 419–432.

Haast J., 1886, "On *Megalapteryx hectori*, a new gigantic species of apterygian birds", Trans. Zool. Soc. Lond., vol. 12, 161–169.

Handley W.D., Chinsamy A., Yates A.M. *et al.*, 2016, "Sexual dimorphism in the late Miocene mihirung *Dromornis stirtoni* (Aves: Dromornithidae) from the Alcoota Local Fauna of central Australia", J. Vertebr. Paleontol., e1180298.

Harris W.K., 1968, "Tasmanian Tertiary and Quaternary microfloras summary report", Dep. Mines South Aust. Geol. Surv. Palaeontol. Sect. Rep., vol. 4017, 5–68.

Hébert E., 1855a, "Note sur le tibia du *Gastornis parisiensis*", C. R. Acad. Sci., vol. 40, 579–582.

Hébert E., 1855b, "Note sur le fémur du *Gastornis parisiensis*", C. R. Acad. Sci., vol. 40, 1214–1217.

Hector J., 1872, "On recent moa remains in New Zealand", Trans. Proc. N. Z. Inst., vol. 4, 110–120.

Hellmund M., 2013, "Reappraisal of the bone inventory of *Gastornis geiselensis* (Fischer, 1978) from the Eocene Geiseltal Fossillagerstatte (Saxony-Anhalt, Germany)", N. Jb. Geol. Paläontol. Abh., vol. 269, 203–220.

Henrici P., 1957, "*Aepyornis*-Eier", Mitt. Naturforsch. Ges. Bern., vol. 14, 135–139.

Holdaway R.N., Worthy T.H., 1997, "A reappraisal of the late Quaternary fossil vertebrates of Pyramid valley swamp, North Canterbury, New Zealand", N. Z. J. Zool., vol. 24, 69–121.

Hooker J.J., 2012, "A new omomyid primate from the earliest Eocene of southern England: First phase of microchoerine evolution", Acta Palaeontol. Pol., vol. 57, 449–462.

Hooker J.J., Collinson M.E., Lawson A.G. *et al.*, 2009, "The Woolwich Formation of Croydon, S. London, UK: a PETM fauna and flora rediscovered", in Crouch E.M., Strong C.P., Hollis C.J. (Eds.) Climatic and Biotic Events of the Paleogene (CBEP 2009), GNS Science Miscellaneous Series 18, extended abstracts, Wellington, pp. 78–81.

Horner J.R., 2000, "Dinosaur reproduction and parenting", Ann. Rev. Earth Planet. Sci., vol. 28, 19–45.

Horner J.R., Padian K. 2004, "Age and growth dynamics of *Tyrannosaurus rex*", Proc. R. Soc. Lond. B, vol. 271, 1875–1880.

Horton D., 1978, "Extinction of the Australian megafauna", Aust. Inst. Aborig. Stud. Newsl., vol. 9.

Hou L.-H., 1980, "New form of the Gastornithidae from Lower Eocene of the Xichuan, Honan", Vertebr. Palasiat., vol. 18, 111–115.

Hoyt D.F., 1979, "Practical methods of estimating volume and fresh weight of bird eggs", The Auk, vol. 96, 73–77.

Hume J.P., Walters M., 2012, "Extinct birds", Poyser, London, 544 p.

Hurlburt G., 1999, "Comparison of body mass estimation techniques, using recent reptiles and the pelycosaur *Edaphosaurus boanerges*", J. Vertebr. Paleontol., vol. 19, 338–350.

Hutton F.W., 1893, "On *Dinornis (?) queenslandiae*", Proc. Linn. Soc. N. S. W., vol. 8, 7–10.

Huynen L., Millar C.D., Scofield R.P. *et al.*, 2003, "Nuclear DNA sequences detect species limits in ancient moa", Nature, vol. 425, 175–178.

Huynen L., Gill B.J., Millar C.D. *et al.*, 2010, "Ancient DNA reveals extreme egg morphology and nesting behavior in New Zealand's extinct moa", PNAS, vol. 107, 16201–16206.

Huynen L., Gill B.J., Doyle A. *et al.*, 2014, "Identification, classification, and growth of moa chicks (Aves: Dinornithiformes) from the genus *Euryapteryx*", PlosOne, vol. 9, e99929. doi:10.1371/journal.pone.0099929

Johnson B.J., Miller G.H., Fogel M.L. *et al.*, 1997, "The determination of late Quaternary paleoenvironments at Equus Cave, South Africa, using stable isotopes and amino acid racemization in ostrich eggshell", Palaeogeogr. Palaeoclimatol. Palaeoecol., vol. 136, 121–137.

Jones W.W., Rinderknecht A., Montenegro F. *et al.*, 2016, "New evidence of terror birds (Cariamae, Phorusrhacidae) from the late Pleistocene of Uruguay", in 9th International Meeting of the Society of Avian Paleontology and Evolution, Diamante Abstracts, p. 16.

Kay R.F., Bruce J. Macfadden B.J. *et al.*, 1998. Revised age of the Salla beds, Bolivia, and its bearing on the age of the Deseadan South American Land Mammal "Age". Journal of Vertebrate Paleontology 18:189–199.

Kennedy B.V., Krouse H.R., 1990, "Isotope fractionation by plants and animals: implications for nutrition research", Can. J. Physiol. Pharmacol., vol. 68, 960–972.

Kérourio P., 1981, "Nouvelles observations sur le mode de nidification et de ponte chez les dinosauriens du Crétacé terminal du Midi de la France", C. R. Soc. géol. Fr., vol. 1, 25–28.

Kerourio P., Aujard C., 1987, "Les œufs d'oiseaux géants du Midi de la France", Monde Minér, vol. 78, 12–16.

Klein N., Sander P.M., Le Lœuff J. *et al.*, 2012, "Modified laminar bone in *Ampelosaurus atacis* and other titanosaurs (Sauropoda): Implications for life history and physiology", PLoS One, vol. 7, 1–17.

Koch P.L., 2007, "Isotopic study of the biology of modern and fossil vertebrates", in Michener R.H., Lajtha K. (Eds.), Stable isotopes in ecology and environmental science Second Edition, Ecological methods and concepts Series. John Wiley & Sons, London, pp. 99–154.

Koch P.L., Fogel M.L., Tuross N., 1994, "Tracing the diets of fossil animals using stable isotopes", in Michener R.H., Lajtha K. (Eds.), Stable isotopes in ecology and environmental science Second Edition, Ecological methods and concepts Series. John Wiley & Sons, London, pp. 63–92.

Köhler M., Moyà-Solà S., 2009, "Physiological and life history strategies of a fossil large mammal in a resource-limited environment", Proc. Natl. Acad. Sci., vol. 106, 20354–20358.

Kohn M.J., 2010, "Carbon isotope compositions of terrestrial C3 plants as indicators of (paleo)ecology and (paleo)climate", Proc. Natl. Acad. Sci., vol. 107, 19691–19695.

Kraglievich L., 1931, "Contribución al conocimentio de la aves fósiles de la época araucoentreriana", Physis, vol. 10, 304–315.

Kraglievich L., 1932, "Una gigantesca ave fósil del Uruguay, *Devincenzia gallinali* n. gen. n. sp., tipo de una nueva familia, Devincenziidae, del Orden Stereornithes", An. Mus. Hist. Nat. Montevideo, vol. 3, 323–355.

Krumbiegel I. 1966, "Die Straussenvögel", A. Ziemsen Verlag, Wittenberg Lutherstadt.

Lamberton C., 1934, "Ratites subfossiles de Madagascar. Les Mullerornithidae", Mém. Académie Malgache, vol. 17, 125–168.

Lambrecht K., 1929, "Ergebnisse der Forschungsreisen Prof. E. Stromers in den Wüsten Aegyptens. V. Tertiäre Wirbeltiere. 4. *Stromeria fajumensis* n. g., n. sp., die kontinentale Stammform der Aepyornithidae, mit einer Übersicht über die fossilen Vögel Madagaskars und Afrikas", Abh. Bayer. Akad. Wiss. Math.-Naturwiss. Abt. N. F., vol. 4, 1–18.

Lambrecht K., 1933, "Handbuch der Palaeornithologie", Gebrüder Borntraeger. Berlin.

Lartet E., 1855, "Note sur le tibia d'oiseau fossile de Meudon", C. R. Acad. Sci. Paris, vol. 40, 582–583.

Last J.T., 1894, "On the bones of the *Aepyornis*, and on the localities and conditions in which they are found", Proc. Zool. Soc. Lond., 1894, 123–129.

Laurent Y., Adnet S., Bourdon E. *et al.*, 2010, "La Borie (Saint-Papoul, Aude) : un gisement exceptionnel dans l'Éocène basal du Sud de la France", Bull. Soc. Hist. Nat. Toulouse, vol. 146, 89–103.

Lavauden L., 1931, "Animaux disparus et légendaires de Madagascar", Rev. Sci. Illus. 69e année, 297–308.

Lazzerini N., Lécuyer, C., Amiot, R. *et al.*, 2016, "Oxygen isotope fractionation between bird eggshell calcite and body water", Sci. Nat. 1–15.

Lécuyer C., 2013, "Water on Earth", Wiley-ISTE, 272 p.

Lee K., Feinstein J., Cracraft J., 1997, "The phylogeny of ratite birds: resolving conflicts between molecular and morphological data sets", in Mindell D.P. (Ed.), Avian molecular evolution and systematics. Academic Press, San Diego, pp. 173–211.

Lee-Thorp J.A., Van Der Merwe N.J., Brain C.K., 1989, "Isotopic evidence for dietary differences between two extinct baboon species from Swartkrans", J. Hum. Evol., vol. 18, 183–189.

Legendre S., Crochet J.Y., Godinot M. *et al.*1991, "Evolution de la diversité des faunes de mammifères d'Europe occidentale au Paléogène (MP 11 to MP 30)", Bull. Soc. géol. Fr., vol. 162, 867–874.

Lemoine V., 1878, "Recherches sur les oiseaux fossiles des terrains tertiaires inférieurs des environs de Reims", part one, Keller, Reims.

Lemoine V., 1881, "Recherches sur les oiseaux fossiles des terrains tertiaires inférieurs des environs de Reims", part two, Matot-Braine, Reims.

Lenz O.K., Wilde V., Mertz D.F. *et al.*, 2015, "New palynology-based astronomical and revised $^{40}Ar/^{39}Ar$ ages for the Eocene maar lake of Messel (Germany)", Int. J. Earth Sci., vol. 104, 873–889.

Linnaeus C., 1758, "Systema naturae per regna tria naturae, secundum classes, ordines, genera, species, cum characteribus, differentiis, synonymis, locis". Laurentius Salvius, Stockholm.

Long J., 2002, "The dinosaur dealers", Allen & Unwin, Crows Nest NSW, Australia.

Long J., Vickers-Rich P., Hirsch K. *et al.*, 1998, "The Cervantes egg: an early Malagasy tourist to Australia", Rec. West. Aust. Mus., vol. 19, 39–46.

Louis P., 1996, "Recherches de mammifères paléogènes dans les départements de l'Aisne et de la Marne pendant la deuxième moitié du vingtième siècle". Palaeovertebrata, Volume jubilaire, D.E. Russell, vol. 25, 83–113.

Lundelius E.J., 1960, "Post-Pleistocene faunal succession in Western Australia and its climatic interpretation", in Report of the International Geological Congress. 21st International Geological Congress, pp. 142–153.

Lydekker R., 1891, "Catalogue of the Fossil Birds in the British Museum", Taylor & Francis, London.

Lydekker R., 1895, "The giant birds of South America", Knowledge, vol. 18, 125–127.

Lydekker R., 1896, "Stereornithes", in Newton A., Gadow H. (Eds.), A Dictionary of Birds. London, pp. 904–908.

MacFadden B.J., Labs-Hochstein J., Hulbert R.C. *et al.*, 2007, "Revised age of the late Neogene terror bird (*Titanis*) in North America during the Great American Interchange", Geology, vol. 35, 123–126.

MacPhail M.K., 1995, "A provisional palynostratigraphic framework for Tertiary organic facies in the Burt Plain, Hale, Ngalia, Santa Teresa, Ti-tree, and Waite Basins", AGSO Wiluraratja Kapi West. Waters Proj. Palynol. Complet. Rep., vol. 1, 1–23.

MacPhail M.K., 1997, "Late Neogene climates in Australia: fossil pollen and spore-based estimates in retrospect and prospect", Aust. J. Bot., vol. 45, 425–464.

MacPhee R.D.E., Burney D.A., Wells N.A., 1985, "Early Holocene chronology and environment of Ambasambazimba, a Malagasy subfossil lemur site", Int. J. Primatol., vol. 6, 463–489.

Mahé J., 1972, "The Malagasy subfossils", in Biogeography and ecology in Madagascar, Junk, B.V., 339–365.

Mahé J., Sourdat M., 1972, "Sur l'extinction des vertébrés subfossiles et l'aridification du climat dans le Sud-Ouest de Madagascar", Bull. Soc. Geol. Fr., vol. 14, 295–309.

Marden L., 1967, "Madagascar: island at the end of the earth", Natl. Geogr. Mag., vol. 132, 443–487.

Marsh O.C., 1894, "A gigantic bird from the Eocene of New Jersey", Am. J. Sci., vol. 48, 344.

Martin H.A., 2006, "Cenozoic climatic change and the development of the arid vegetation in Australia", J. Arid Environ., vol. 66, 533–563.

Martin L.D., 1992, "The status of the late Paleocene birds *Gastornis* and *Remiornis*", in Campbell K.E. (Ed.), Papers in Avian Paleontology Honoring Pierce Brodkorb, Nat. Hist. Museum of Los Angeles County Science Series. Los Angeles, pp. 97–108.

Martini E., 1961, "Neue Funde von fossilen Eiern in der Provence. Gelege von Straussenvögeln aus dem Alttertiär ?", Umschau, vol. 10, 311–312.

Matthew W.D., Granger W., 1917, "The skeleton of *Diatryma*, a gigantic bird from the Lower Eocene of Wyoming", Bull. Am. Mus. Nat. Hist., vol. 37, 307–326.

Mayr G., 2009, "Paleogene fossil birds", Springer, Berlin, 262 p.

Mayr G., 2011, "Cenozoic mystery birds – on the phylogenetic affinities of bony-toothed birds (Pelagornithidae)", Zool. Scr., vol. 40, 448–467.

Mayr G., 2016, "Avian evolution. The fossil record of birds and its paleobiological significance", Wiley-Blackwell, Chichester.

Megirian D., Murray P., Wells R., 1996, "The Late Miocene Ongeva Local Fauna of Central Australia", Beagle Rec. Mus. Art Galleries North. Territ., vol. 13, 9–38.

Megirian D., Prideaux G.J., Murray P.F. *et al.*, 2010, "An Australian land mammal age biochronological scheme", Paleobiology, vol. 36, 658–671.

Mertz D.F., Renne P.R., 2005, "A numerical age for the Messel fossil deposit (UNESCO World Heritage Site) derived from $^{40}A/^{39}A$ dating on a basaltic rock fragment", Cour. Forschungsinstitut Senckenberg, vol. 255, 67–75.

Miller A.H., 1963, "Fossil ratite birds of the late Tertiary of South Australia", Rec. South Aust. Mus., vol. 14, 413–420.

Miller G.H., Magee J.W., Johnson, B.J. *et al.*, 1999, "Pleistocene extinction of *Genyornis newtoni*: Human impact on Australian megafauna", Science, vol. 283, 205–208.

Miller K.G., 2009, "Sea level change, last 250 million years", in Encyclopedia of Paleoclimatology and ancient environments, Springer, Berlin, pp. 879–887.

Milne-Edwards A., 1863, "Mémoire sur la distribution géographique des oiseaux fossiles et description de quelques espèces nouvelles", Ann. Sci. Nat. Zool., vol. 20, 133–176.

Milne-Edwards A., 1867, "Recherches anatomiques et paléontologiques pour servir à l'histoire des oiseaux fossiles de la France" (vol. 1), Masson, Paris.

Milne-Edwards A., 1872, "Oiseaux fossiles", in Orbigny C. d' (Ed.), Dictionnaire Universel d'Histoire Naturelle. Paris, pp. 670–719.

Milne-Edwards A., 1873, "Lettre accompagnant l'envoi des "Recherches sur la faune ornithologique éteinte des îles Mascareignes et de Madagascar et parlant des ossements de *l'Aepyornis* découvert par Grandidier"", C. R. Acad. Sci., vol. 77, 1337.

Milne-Edwards A., Grandidier A., 1869, "Nouvelles observations sur les caractères zoologiques et les affinités naturelles de l'*Aepyornis* de Madagascar", Ann. Sci. Nat., vol. 7, 85–114.

Milne-Edwards A., Grandidier A., 1894, "Observations sur les *Aepyornis* de Madagascar", C. R. Acad. Sci., vol. 118, 122–127.

Mitchell K.J., Llamas B., Soubrier J. *et al.*, 2014, "Ancient DNA reveals elephant birds and kiwi are sister taxa and clarifies ratite bird evolution", Science, vol. 344, 898–900.

Mitchell T.L., 1839, "Three expeditions into the interior of the Eastern Australia", Moore. London.

Mlíkovský J., 2002, "Cenozoic birds of the world. Part 1: Europe", NINOX Press. Praha.

Mlíkovský J., 2003, "Eggs of extinct aepyornithids (Aves: Aepyornithidae) of Madagascar: size and taxonomic identity", Sylvia, vol. 39, 133–138.

Monnier L., 1913, "Paléontologie de Madagascar. VII. Les *Aepyornis*". Annales de Paléontologie, vol. 8, pp. 125–172.

Moreno F.P., 1889, "Breve reseña de los progresos del Museo La Plata, durante el segundo semestre de 1888", Bol. Mus. Plata, vol. 3, 5–44.

Moreno F.P., Mercerat A., 1891, "Catalogue des oiseaux fossiles de la République Argentine conservés au Musée de La Plata", Anales del Museo de La Plata, vol. 1, 7–71.

Mosbrugger V., Utescher T., Dilcher D.L., 2005, "Cenozoic continental climatic evolution of Central Europe", Proc. Natl. Acad. Sci. U. S. A., vol. 102, 14964–14969.

Mourer-Chauviré C., Bourdon E., 2015, "The *Gastornis* (Aves, Gastornithidae) from the Late Paleocene of Louvois (Marne, France)", Swiss J. Palaeontol., 1–15.

Mourer-Chauviré C., Tabuce R., Mahboubi M. *et al.*, 2011, "A Phororhacoid bird from the Eocene of Africa", Naturwissenschaften, vol. 98, 815–823.

Murray P.F., Megirian D., 1992, "Continuity and contrast in Middle Miocene vertebrate communities from the Northern Territory", Beagle Rec. North. Territ. Mus. Arts Sci., vol. 9, 195–218.

Murray P.F., Megirian D., 1998, "The skull of dormornithid birds: Anatomical evidence for their relationship to Anseriformes", Rec. South Aust. Mus., vol. 31, 51–97.

Murray P.F., Vickers-Rich P., 2004, "Magnificent Mihirungs - The Colossal Flightless Birds of the Australian Dreamtime", Indiana University Press. Bloomington, 416 p.

Murray P.F., Megirian D., Wells R., 1993, "*Kolopsis yperus* sp. nov. (Zygomaturinae, Marsupialia) from the Ongeva Local Fauna: new evidence for the age of the Alcoota fossil beds of Central Australia", Beagle Rec. North. Territ. Mus. Arts Sci., vol. 10, 155–172.

Mustoe G.E., Tucker D.S., Kemplin K.L., 2012, "Giant Eocene bird footprints from northwest Washington, USA", Palaeontology, vol. 55, 1293–1305.

Nasif N.L., Esteban G.I., Ortiz P.E., 2009, "Novedoso hallazgo de egagrópilas en el Mioceno tardío, Formación Andalhuala, provincia de Catamarca, Argentina", Ser. Correl. Geol., vol. 25, 105–114.

Needham J., 1931, "Chemical Embryology", Cambridge University Press. Cambridge UK.

Newsome A.D., Rochow K.A., 1964, "Vertebrate fossils from Tertiary sediments of Central Australia", Aust. J. Sci., vol. 26, 352.

Newton E.T., 1884, "Ornithology", Encycl. Br., vol. 18.

Newton E.T., 1885a, "*Gastornis klaasseni* Newton, a gigantic bird from the Lower Eocene of Croydon", Geol. Mag., vol. 3, 362–363.

Newton E.T., 1885b, "On the remains of a gigantic species of bird from the Lower Eocene beds near Croydon", Proc. Zool. Soc. Lond., vol. 53, 445–446.

Newton E.T., 1886, "On the remains of a gigantic species of bird (*Gastornis klaasseni*, n. sp.) from the Lower Eocene beds near Croydon", Trans. Zool. Soc. Lond., vol. 12, 143–160.

Nguyen J.M.T., Boles W.E., Hand S.J., 2010, "New material of *Barawertornis tedfordi*, a Dromornithid from the Oligo-Miocene of Australia, and its phylogenetic implications", Rec. Aust. Mus., vol. 62, 45–60.

Noriega J.I., Agnolín F.L., 2008, "El registro paleontológico de las Aves del" Mesopotamiense" (Formación Ituzaingó ; Mioceno tardío-Plioceno) de la provincia de Entre Ríos, Argentina", INSUGEO Miscelánea, vol. 17, 271–290.

O'Leary M.H., 1988, "Carbon isotopes in photosynthesis", Bioscience, vol. 38, 328–336.

Oliver W.R.B., 1949, "The moas of New Zealand and Australia", Dom. Mus. Bull., vol. 15, 1–205.

Osborn H.F., 1934, "Aristogenesis, the creative principle in the origin of species", Am. Nat., vol. 68, 193–235.

Owen R., 1839, "Exhibition of a bone of an unknown struthious bird from New Zealand", Proc. Zool. Soc. Lond., vol. 7, 169–171.

Owen R., 1843a, "On *Dinornis*, an extinct genus of tridactyle struthious birds, with descriptions of portions of the skeleton of five species which formerly existed in New Zealand", Trans. Zool. Soc. London, vol. 3, 235–275.

Owen R., 1843b, "On *Dinornis novaezealandiae*", Proc. Zool. Soc. of London. pp. 8–10.

Owen R., 1846, "Memoir on the *Dinornis*, Part II", Proc. Zool. Soc. Lond., vol. 14, 46–49.

Owen R., 1848, "On the remains of the gigantic and presumed extinct wingless or terrestrial birds of New Zealand (*Dinornis* and *Palapteryx*), with indications of two other genera (*Notornis* and *Nestor*)", Proc. Zool. Soc. Lond., vol. 16, 1–11.

Owen R., 1856a, "On the affinities of a large extinct bird (*Gastornis parisiensis*, Hébert) indicated by a fossil femur and tibia discovered in the lowest Eocene formation near Paris", Q. J. Geol. Soc. Lond., vol. 12, 204–217.

Owen R., 1856b, "On *Dinornis* (part VII): containing a description of the bones of the leg and foot of the *Dinornis elephantopus*, Owen", Proc. Zool. Soc. London, vol. 24, 54–61.

Owen R., 1872, "Part 19 of Owen's memoir on *Dinornis*, read June 4, 1872", Proc. Zool. Soc. Lond., 682–683.

Owen R., 1874, "On *Dinornis* (part XIX): containing a description of a femur indicative of a new genus of large wingless birds (*Dromornis australis*, Owen) from a post-Tertiary deposit in Queelsland, Australia", Trans. Zool. Soc. Lond., vol. 8, 381–384.

Owen R., 1879a, "Memoirs on the Extinct Wingless Birds of New Zealand, with an Appendix on those of England, Australia, Newfoundland, Mauritius, and Rodriguez" (2 vol.). John van Voorst, London.

Owen R., 1879b, "On *Dinornis*: Containing a restoration of the skeleton of *Dinornis maximus* (Owen), with an appendix on additional evidence of the genus *Dromornis* in Australia", Trans. Zool. Soc. Lond., vol. 10, 147–188.

Owen R., 1883, "On *Dinornis* (part XXIV): Containing a description of the head and feet, with their dried integuments, of an individual of the species *Dinornis didinus*", Trans. Zool. Soc. Lond., vol. 11, 257–261.

Paik I.S., Huh M., Kim H.J., 2004, "Dinosaur egg-bearing deposits (Upper Cretaceous) of Boseong, Korea: occurrence, palaeoenvironments, taphonomy, and preservation", Palaeogeogr. Palaeoclimatol. Palaeoecol., vol. 205, 155–168.

Parker Pearson M., Godden K., Ramilisonina, R. *et al.*, 1995, "The Central Androy Survey" (No. 3), University of Sheffield and Musée d'Art et d'Archéologie, Antananarivo.

Passey B.H., Robinson T.F., Ayliffe L.K. *et al.*, 2005, "Carbon isotope fractionation between diet, breath CO_2, and bioapatite in different mammals", J. Archaeol. Sci., vol. 32, 1459–1470.

Patterson B., 1941, "A new phororhacoid bird from the Deseado formation of Patagonia", Geological Series, Field Museum of Natural History, vol. 8, 49–54.

Patterson B., Kraglievich J.L., 1960, "Systemática y nomenclatura de las aves fororracoideas del Plioceno argentino", Pub. Mus. Mun. Cienc. Nat. Tradicion Mar. del Plata, vol. 1, 3–49.

Patterson J., Lockley M.G., 2004, "A probable *Diatryma* track from the Eocene of Washington: an intriguing case of controversy and skepticism", Ichnos, vol. 11, 341–347.

Penner M.M., 1983, "Contribution à l'étude de la microstructure des coquilles d'œufs de dinosaures du Crétacé supérieur dans le bassin d'Aix-en-Provence : Application biostratigraphique". PhD thesis, Université Pierre et Marie Curie, Paris.

Perkins M.E., Fleagle J.G., Heizler M.T. *et al.*, 2012, "Tephrochronology of the Miocene Santa Cruz and Pinturas Formations, Argentina", in Vizcaíno S.F., Kay R.F., Bargo (Eds.), Early Miocene Paleobiology in Patagonia: High-Latitude Paleocommunities of the Santa Cruz Formation Cambridge University Press, Cambridge, pp. 41–48.

Perry G.L.W., Wheeler A.B., Wood J.R. *et al.*, 2014, "A high-precision chronology for the rapid extinction of New Zealand moa (Aves, Dinornithiformes)", Quat. Sc. Rev., vol. 105, 126–135.

Phillips M.J., Gibb G.C., Crimp E.A. *et al.*, 2010, "Tinamous and moa flock together: mitochondrial genome sequence analysis reveals independent losses of flight among ratites", Syst. Biol., vol. 59, 90–107.

Plane M., Gatehouse C.G., 1968, "A new vertebrate fauna from the Tertiary of northern Australia", Aust. J. Sci., vol. 30, 272–273.

Pouech J.J., 1859, "Mémoire sur les terrains tertiaires de l'Ariège rapportés à une coupe transversale menée de Fossat à Aillières, passant par le Mas d'Azil, et projetée sur le méridien de ce lieu", Bull. Soc. Geol. Fr., vol. 16, 381–411.

Prévost C., 1855a, "Annonce de la découverte d'un oiseau fossile de taille gigantesque, trouvé à la partie inférieure de l'argile plastique du terrain parisien", C. R. Acad. Sci. Paris, vol. 40, 554–557.

Prévost C., 1855b, "Nouveaux documents sur le gisement du *Gastornis parisiensis* et considérations générales sur les vestiges laissées par des oiseaux dans les terrains des divers âges", C. R. Acad. Sci., Paris, vol. 40, 616–619.

Prichinello K.A., 1971, "Earliest Eocene mammalian fossils from the Laramie Basin of southeast Wyoming", Rocky Mt. Geol., vol. 10, 73–87.

Pycraft W.P., 1900, "On the morphology and phylogeny of the Palaeognathae (Ratitae and Crypturi) and Neognathae (Carinatae)", Trans. Zool. Soc. Lond., vol. 15, 149–290.

Rahn H., Ar A., 1974, "The avian egg: incubation time and water loss", The Condor, vol. 76, 147–152.

Rahn H., Paganelli C.V., Ar A., 1975, "Relation of avian egg weight to body weight", The Auk, vol. 92, 750–765.

Rahn H., Sotherland P.R., Paganelli C.V., 1985, "Interrelationships between egg mass and adult body mass and metabolism among passerine birds", J. Ornithol., vol. 126, 263–271.

Rakotozafy L.M., Goodman S.M., 2005, "Contribution à l'étude zooarchéologique de la région du Sud-ouest et extrême Sud de Madagascar sur la base des collections de l'ICMAA de l'Université d'Antananarivo", Talaha, 14–15.

Rasmussen D.T., Simons E.L., Hertel P. *et al.*, 2001, "Hindlimb of a giant terrestrial bird from the upper Eocene, Fayum, Egypt", Palaeontology, vol. 44, 325–337.

Rawlence N.J., Wood J.R., Armstrong K.A. *et al.*, 2009, "DNA content and distribution in ancient feathers and potential to reconstruct the plumage of extinct avian taxa", Proc. Roy. Soc. Lond. B, vol. 276, 3395–3402.

Rayfield E.J., 2007, "Finite element analysis and understanding the biomechanics and evolution of living and fossil organisms", Annu. Rev. Earth Planet Sci., vol. 35, 541–576.

Reichenbach H.G.L., 1852, "Avium systema naturale. Das natürliche System der Vögel", Expedition der vollständigsten Naturgeschichte. Dresden & Leipzig.

Rey K., Amiot R., Lécuyer C. *et al.* 2013, "Late Miocene climatic and environmental variations in northern Greece inferred from stable isotope compositions ($\delta^{18}O$, $\delta^{13}C$) of equid teeth apatite", Palaeogeogr. Palaeoclimatol. Palaeoecol., vol. 388, 48–57.

Rich P.V., 1979, "The Dromornithidae, an extinct family of large ground birds endemic to Australia", Bull. Bur. Miner. Resour. Geol. Geophys., vol. 184, 1–188.

Rich P.V., Gill E., 1976, "Possible dromornithid footprints from Pleistocene dune sands of southern Victoria, Australia", Emu, vol. 76, 221–223.

Rich P.V., Green R.H., 1974, "Bird footprint at South Mt. Cameron, Tasmania", Emu, vol. 74, 245–248.

Rose K.D., 1981, "The Clarkforkian Land-Mammal Age and mammalian faunal composition across the Paleocene-Eocene boundary", Univ. Mich. Pap. Paleontol., 1–197.

Rovereto C., 1914, "Los estratos araucanos y sus fósiles", An. Mus. Nac. Hist. Nat. B. Aires, vol. 25, 1–247.

Rowley G.D., 1867, "On the egg of *Aepyornis*, the colossal bird of Madagascar", Proc. Zool. Soc. Lond., 892–895.

Scarlett R., 1969, "On the alleged Queensland moa, *Dinornis queenslandiae* DeVis", Mem. Qld. Mus., vol. 15, 207–212.

Schaub S., 1929, "Über eocäne Ratitenrste in der osteologischen Sammlung des Basler Museums", Verhandl. Naturforsch. Ges. Basel, vol. 40, 588–598.

Schaub S., 1940, "Ein Ratitenbecken aus dem Bohnerz von Egerkingen", Eclogae Geol. Helvetiae, vol. 32, 274–284.

Schmidt-Nielsen K., 1985, "Scaling: why is animal size so important?", Cambridge University Press, London.

Schmitz B., Pujalte V., 2007, "Abrupt increase in seasonal extreme precipitation at the Paleocene-Eocene boundary", Geology, vol. 35, 215–218.

Shockey B.J., Gismondi R.S., Gans P. *et al.*, 2009, "Paleontology and Geochronology of the Deseadan (Late Oligocene) of Moquegua, Peru", Am. Mus. Novitates, 3668.

Shufeldt R.W., 1909, "Osteology of birds", N. Y. State Mus., vol. 130, 1–381.

Sibley C.G., Ahlquist J.E., Monroe Jr B.L., 1988, "A classification of the living birds of the world based on DNA-DNA hybridization studies", The Auk, vol. 105, 409–423.

Sinclair W.J., 1928a, "*Omorhamphus*, a new flightless bird from the Lower Eocene of Wyoming", Proc. Am. Philos. Soc., 51–65.

Sinclair W.J., 1928b, "*Omorhamphus*, a new fossil flightless bird", The Auk, vol. 45, 522–523.

Sinclair W.J., Farr M.S., 1932, "Aves of the Santa Cruz Beds", Rep Princet. Univ Exped Patagon. 1896–1899, vol. 7, 157–191.

Sluijs A., Brinkhuis H., Crouch E.M. *et al.*, 2008, "Eustatic variations during the Paleocene-Eocene greenhouse world", Paleoceanography, vol. 23, doi: 10.1029/2008PA001615.

Smektala F., Buffetaut E., Deconinck J.F., 2014, "Rivers as repositories for fossil vertebrates: A case study from the Upper Cretaceous of southern France", Proc. Geol. Assoc., vol. 125, 567–577.

Smith J.M., Savage R.J., 1956, "Some locomotory adaptations in mammals", J. Linn. Soc. Lond. Zool., vol. 42, 603–622.

Smith T., Quesnel F., De Plöeg G. *et al.* 2014, "First Clarkforkian equivalent Land Mammal Age in the Latest Paleocene basal Sparnacian facies of Europe: fauna, flora, paleoenvironment and (Bio) stratigraphy", PLoS One, vol. 9, e86229.

Solé F., Falconnet J., Vidalenc D., 2015, "New fossil Hyaenodonta (Mammalia, Placentalia) from the Ypresian and Lutetian of France and the evolution of the Proviverrinae in southern Europe", Palaeontology, vol. 58, 1049–1072.

Spry A., Banks M.R., 1962, "The geology of Tasmania", J. Geol. Soc. Aust., vol. 9, 107–362.

Stein B.R., Casinos A., 1997, "What is a cursorial mammal?", J. Zool., vol. 242, 185–192.

Stejneger L., 1885, "Natural history of birds" in Kingsley J.S. (Ed.), The Standard Natural History, Boston, Cassino, pp. 1–195.

Stidham T.A., Eberle J.J., 2016, "The palaeobiology of high latitude birds from the early Eocene greenhouse of Ellesmere Island, Arctic Canada", Sci. Rep., vol. 6.

Stirling E.C., 1896, "The newly discovered extinct gigantic bird of South Australia", Ibis, vol. 7, 593.

Stirling E.C., Zietz A.H.C., 1896, "Preliminary notes on *Genyornis newtoni*: a new genus and species of fossil struthious bird found at Lake Callabonna, South Australia", Trans. R. Soc. S. Aust., vol. 20, 171–211.

Stirling E.C., Zietz A.H., 1900, "Fossil remains of Lake Callabonna. Part II.1. *Genyornis newtoni*: A new genus and species of struthious bird", Mem. R. Soc. S. Aust., vol. 1, 41–80.

Stirling E.C., Zietz A.H.C., 1905, "Fossil remains of Lake Callabonna. Part III. Description of the vertebrae of *Genyornis newtoni*", Mem. R. Soc. S. Aust., vol. 1, 81–110.

Stirling E.C., Zietz A.H., 1913, "Fossil remains of Lake Callabonna. Part IV: Description of some remains of *Genyornis newtoni*", Mem. R. Soc. S. Aust., vol. 1, 111–126.

Stirton R.A., Tedford R.H., Woodburne M.O., 1967, "A new Tertiary formation and fauna from the Tirari Desert, South Australia", Rec. South Aust. Mus., vol. 15, 427–462.

Storer R., 1960, "Adaptive radiation in birds", in: Marshall A.J. (Ed.) Biology and comparative physiology of birds. A. J. Marshall, Academic Press, New York & London, pp. 15–55.

Sudre J., Legendre S., 1992, "Ungulates from Paleogene of western Europe: relationship between their evolution and environmental changes during that period", in Spitz F., Janeau G., Gonzalez G., Aulagnier S. (Eds.), Ongules Ungulates 91; International Symposium, Toulouse, pp. 15–25.

Tambussi C.P., 2011, "Palaeoenvironmental and faunal inferences based on the avian fossil record of Patagonia and Pampa: what works and what does not", Biol. J. Linn. Soc., vol. 103, 458–474.

Tambussi C.P., Degrange F.J., 2013, "South American and Antarctic continental Cenozoic birds. Paleobiogeographic affinities and disparities", Springer, Dordrecht.

Tambussi C.P., Ubilla M., Perea D., 1999, "The youngest large carnassial bird (Phorusrhacidae, Phorusrhacinae) from South America (Pliocene–Early Pleistocene of Uruguay)", J. Vertebr. Paleontol., vol. 19, 404–406.

Tattersall I., 1973, "A note on the age of the subfossil site of Ampasambazimba, Miarinarivo Province, Malagasy Republic", Am. Mus. Novit., vol. 2520, 1–6.

Tedford R.H., 1966, "A review of the macropod genus *Sthenurus*", Univ. Calif. Publ. Geol. Sci., vol. 57.

Tedford R.H., 1967, "The fossil Macropodidae from Lake Menindee, New South Wales". Univ. Calif. Publ. Geol. Sci.

Tedford R.H., 1968, "Fossil mammal remains from the Tertiary Carl Creek Limestone, north-western Queensland", Bull. Bur. Miner. Resour. Geol. Geophys., vol. 92, 217–237.

Tedford R.H., 1971, "Quaternary vertebrates and environments, Lake Callabonna, South Australia", in 43rd Cong. Australian and New Zealand Association Adv. Sci., Sect. 3. pp. 7–8.

Tedford R.H., 1984, "The diprotodons of Lake Callabonna", Australian Natural History, vol. 17, 349–354.

Tennyson A.J., Worthy T.H., Jones C.M. *et al.*, 2010, "Moa's Ark: Miocene fossils reveal the great antiquity of moa (Aves: Dinornithiformes) in Zealandia", Rec. Aust. Mus., vol. 62, 105–114.

Teodoridis V., Kvaček Z., Zhu H. *et al.*, 2012, "Environmental analysis of the mid-latitudinal European Eocene sites of plant macrofossils and their possible analogues in East Asia", Palaeogeogr. Palaeoclimatol. Palaeoecol., vol. 333, 40–58.

Tonni E.P., Noriega J.I., 1998, "Los cóndores (Ciconiformes, Vulturidae) de la región pampeana de la Argentina durante el Cenozoico tardío : distribución, interacciones y extinciones", Ameghiniana, vol. 32, 141–150.

Tonni E.P., Tambussi C.P., 1986, "Las Aves del Cenozoico de la República Argentina", IV Congreso Argentino de Paleontología y Bioestratigrafía, vol. 2, 131–142.

Tonni E.P., Tambussi C.P., 1988, "Un nuevo Psilopterinae (Aves: Ralliformes) del Mioceno tardio de la provincia de Buenos Aires, Republica Argentina", Ameghiniana, vol. 25, 155–160.

Tooher J., 1998, "Jamie and the elephant egg", Aust. Prop. Law, vol. 6, 117–143.

Torcelli A.J., 1936, "Obras completas y correspondencia cientifica de Florentino Ameghino", La Plata, Taller de Impresiones Oficiales, vol. 21.

Touraine F., 1960, "Œufs d'oiseaux de très grande taille dans l'Éocène inférieur de Provence", Bull. Soc. géol. Fr., vol. 2.

Touraine F., 1961, "Les œufs d'oiseaux éocènes et la stratigraphie du Cengle (B-du-Rh)", C. R. Soc. géol. Fr., vol. 4, 90–91.

Touraine F., 1962, "Le synclinal danien-éocène de la Mourotte (Var)", Bull. Soc. géol. Fr., vol. 4.

Touraine F., 1963, "Quelques précisions nouvelles sur la série fluvio-lacustre de Sillans (Var)", C. R. Soc. géol. Fr., vol. 167.

Touraine F., 1967, "Les rapports de l'Oligocène avec les structures N-S du Nord-Varois", Bull. Soc. géol. Fr., vol. 7, 530–536.

Touraine F., 1978, "Le fossé de Montmeyan (Var)", Bull. BRGM, vol. 1, 89–109.

Tovondrafale T., Razakamanana T., Hiroko K. *et al.*, 2014, "Paleoecological analysis of elephant bird (Aepyornithidae) remains from the Late Pleistocene and Holocene formations of southern Madagascar", Malagasy Nat., vol. 8, 1–13.

Travouillon K.J., Archer M., Hand S.J. *et al.*, 2006, "Multivariate analyses of Cenozoic mammalian faunas from Riversleigh, north-western Queensland", Alcheringa Spec. Issue, vol. 1, 323–349.

Tresize P., 1971, "Rock Art of South-east Cape York", Australian Institute of Aboriginal Studies. Canberra.

Trouessart E., 1895, "Les oiseaux géants de la Patagonie australe", La Nature, vol. 23, 87–91.

Troxell E.L., 1931, "*Diatryma*, a colossal heron", Am. J. Sci., vol. 22, 18–34.

Turvey S.T., Green O.R., Holdaway R.N., 2005, "Cortical growth marks reveal extended juvenile development in New Zealand moa", Nature, vol. 435, 941–943.

Turvey S.T., Holdaway R.N., 2005, "Postnatal ontogeny, population structure, and extinction of the Giant Moa *Dinornis*", J Morphol, vol. 265, 70–86.

Utescher T., Mosbrugger V., Ivanov D. *et al.*, 2009, "Present-day climatic equivalents of European Cenozoic climates", Earth Planet. Sci. Lett., vol. 284, 544–552.

Valenciennes A., 1854, "Note sur l'*Aepyornis*", C. R. Acad. Sci., vol. 39, 837–838.

Valenciennes A., 1855, "Remarques à l'occasion d'une communication de M. Lartet sur le tibia d'un oiseau fossile trouvé à Meudon", C. R. Acad. Sci., vol. 40, 583–584.

Vandenberghe N., Hilgen F.J., Speijer R.P. *et al.* 2012, "The Palaeogene period". In Gradstein F.M., Ogg J.G., Schmitz M., Ogg G. (Eds.), The Geologic Time Scale 2012. Elsevier, Amsterdam, pp. 855–921.

Verheyen R., 1957, "Contribution au démembrement de l'ordo artificiel des Gruiformes. II. Les Cariamiformes", Bull. Inst. Roy. Sc. Nat. Belgique, vol. 33, 1–7.

Vezzosi R.I., 2012a, "First record of *Procariama simplex* Rovereto, 1914 (Phorusrhacidae, Psilopterinae) in the Cerro Azul Formation (upper Miocene) of La Pampa Province; remarks on its anatomy, palaeogeography and chronological range", Alcheringa, vol. 36, 157–169.

Vezzosi R.I., 2012b, "Tamaño y estimacion de la masa corporal en *Procariama simplex* Rovereto, 1914 (Aves: Phorusrhacidae: Psilopterinae)", Ameghiniana, vol. 49, 401–408.

Vickers-Rich P., Rich L.S., Rich T.H., 1997, "Australia's Lost World: A history of Australia's backboned animals", Kangaroo Press-the Monash Science Centre. Kenthurst (Sydney).

Von Schirnding Y., Van Der Merwe N.J., Vogel J.C., 1982, "Influence of diet and age on carbon isotope ratios in ostrich eggshell", Archaeometry, vol. 24, 3–20.

Wagler J.G., 1831, "Einige Mittheilungen über Thiere Mexicos", Isis Von Oken, vol. 5, 510–535.

Webb M., 1957, "The ontogeny of the cranial bones, cranial peripheral and cranial parasympathetic nerves, together with a study of the visceral muscles of *Struthio*", Acta Zool.,vol. 38, 81–203.

Weigelt J., 1939, "Die Aufdeckung der bisher ältesten tertiären Säugetierfauna Deutschlands", Nova Acta Leopoldina, vol. 7, 515–528.

West R.M., Dawson M.R., 1978, "Vertebrate paleontology and the Cenozoic history of the North Atlantic region", Polarforschung, vol. 48, 103–119.

Wetmore A., 1967, "Re-creating Madagascar's giant extinct bird", Natl. Geogr. Magzine, vol. 132, 488–493.

Wetmore A., Miller W., 1926, "The revised classification for the fourth edition of the A.O.U. Check-list", Auk, vol. 43, 337–346.

Williams D.L.G., 1981, "*Genyornis* eggshell (Dromornithidae; Aves) from the Late Pleistocene of South Australia", Alcheringa, vol. 5, 133–140.

Williams D.L.G., Vickers-Rich P., 1992, "Giant fossil egg fragment from the Tertiary of Australia", Contrib. Sci. Nat. Hist. Mus. Los Angel. Ctry., vol. 26, 375–378.

Wilmshurst J.M., Hunt T.L., Lipo C.P. *et al.*, 2011, "High-precision radiocarbon dating shows recent and rapid initial human colonization of East Polynesia", PNAS, vol. 108, 1815–1820.

Wiman C., 1935, "Über *Aepyornithes*", Nova Acta Regiae Soc. Sci. Ups., vol. 9, 1–57.

Wiman C., Edinger T., 1940, "Sur les crânes et les encéphales d'*Aepyornis* et de *Mullerornis*", Bull. Académie Malgache, vol. 23, 1–47.

Wings O., 2004, "Identification, distribution, and function of gastroliths in dinosaurs and extant birds with emphasis on ostriches (*Struthio camelus*)". Unpublished PhD thesis, University of Bonn, Bonn.

Witmer L.M., Rose K.D., 1991, "Biomechanics of the jaw apparatus of the gigantic Eocene bird *Diatryma*: implications for diet and mode of life", Paleobiology, vol. 17, 95–120.

Wood J.R., 2008, "Moa (Aves: Dinornithiformes) nesting material from rockshelters in the semi-arid interior of South Island, New Zealand", J. Roy. Soc. N. Z., vol. 38, 115–129.

Wood J.R., Wilmshurst J.M., 2014, "Late Quaternary terrestrial vertebrate coprolites from New Zealand", Quat. Sci. Rev., vol. 98, 33–44.

Wood J.R., Rawlence N.J., Rogers G.M. *et al.*, 2008, "Coprolite deposits reveal the diet and ecology of the extinct New Zealand megaherbivore moa (Aves, Dinornithiformes)", Quat. Sci. Rev., vol. 27, 2593–2602.

Wood J.R., Wilmshurst J.M., Richardson S.J. *et al.*, 2013, "Resolving lost herbivore community structure using coprolites of four sympatric moa species (Aves: Dinornithiformes)", PNAS, vol. 110, 16910–16915.

Woodburne M.O., 1967, "The Alcoota Fauna, central Australia", Bull. Bur. Miner. Resour. Geol. Geophys., vol. 87, 1–187.

Woods J.E.T., 1866, "Report of the geology and mineralogy of the South-eastern district of the colony of South Australia", Cox, Government Printer, Adelaide: WC.

Woods J.E.T., 1882, "Physical structure and geology of Australia", Proc. Linn. Soc. New Wales, vol. 7, 371–389.

Woods J.T., 1960, "Fossiliferous fluviatile and cave deposits", in The Geology of Queensland, Jour. Geol. Soc. Aust. pp. 393–403.

Worth B., 1940, "Egg volumes and incubation periods", The Auk, vol. 57, 44–60.

Worthy T.H., 2009, "Descriptions and phylogenetic relationships of two new genera and four new species of Oligo-Miocene waterfowl (Aves: Anatidae) from Australia", Zool. J. Linn. Soc., vol. 156, 411–454.

Worthy T.H., Holdaway R.N., 2002, "The lost world of the moa: prehistoric life of New Zealand", Bloomington & Indianapolis. Indiana University Press.

Worthy T.H., Scofield R.P., 2012, "Twenty-first century advances in knowledge of the biology of moa (Aves: Dinornithiformes): a new morphological analysis and moa diagnoses revised", N. Z. J. Zool., vol. 39, 87–153.

Worthy T.H., Yates A., 2015, "Connecting the thigh and foot: resolving the association of post-cranial elements in the species of *Ilbandornis* (Aves: Dromornithidae)", Alcheringa, vol. 39, 407–427.

Worthy T.H., Edwards A.R., Millener P.R., 1991, "The fossil record of moas (Aves: Dinornithiformes) older than the Otira (last) Glaciation", J. R. Soc. N. Z., vol. 21, 101–118.

Worthy T.H., Tennyson A.J.D., Jones C. *et al.*, 2007, "Miocene waterfowl and other birds from central Otago, New Zealand", J. Syst. Palaeont., vol. 5, 1–39.

Worthy T.H., Hand S.J., Archer M., 2014, "Phylogenetic relationships of the Australian Oligo-Miocene ratite *Emuarius gidju* (Casuariidae)", Integr. Zool., vol. 9, 148–166.

Worthy T.H., Handley W.D., Archer M. *et al.*, 2016, "The extinct flightless mihirungs (Aves, Dromornithidae): cranial anatomy, a new species, and assessment of Oligo-Miocene lineage diversity", J. Vertebr. Paleontol., vol. 36, e1031345.

Wroe S., 1999, "The bird from hell?", Nat. Aust., 56–63.

Zachos J.C., Dickens G.R., Zeebe R.E., 2008, "An early Cenozoic perspective on greenhouse warming and carbon-cycle dynamics", Nature, vol. 451, 279–283.

Zheng X., Zhou Z., Wang X. *et al.*, 2013, "Hind wings in basal birds and the evolution of leg feathers", Science, vol. 339, 1309–1312.

Zinoviev A.V., 2013, "Notes on pelvic and hindlimb myology and syndesmology of *Emeus crassus* and *Dinornis robustus* (Aves: Dinornithiformes)", in Göhlich U.B., Kroh A. (Eds.), Proc. 8th Internat. Meet. Soc. Avian Pal. Evol. Naturhistorisches Museum Wien, pp. 253–278.

Zittel K.A., 1902, "Text-book of Paleontology", Macmillan and Company, London.

Index

D, E, F

G, H, I

K, L

M, N, O

P, Q, R

S, T, U

V, W

Kolibri 360 GmbH
Ferdinand-Jühlke-Straße 7
99095 Erfurt, Deutschland
produktsicherheit@kolibri360.de